Springer Tracts in Advanc
Volume 12

Editors: Bruno Siciliano · Oussama Khati

Springe
Berlin
Heidelberg
New York
Hong Kong
London
Milan
Paris
Tokyo

K. Iagnemma · S. Dubowsky

Mobile Robots
in Rough Terrain

Estimation, Motion Planning, and Control with Application to Planetary Rovers

With 58 Figures

Springer

Professor Bruno Siciliano, Dipartimento di Informatica e Sistemistica, Università degli Studi di Napoli Federico II, Via Claudio 21, 80125 Napoli, Italy, email: siciliano@unina.it

Professor Oussama Khatib, Robotics Laboratory, Department of Computer Science, Stanford University, Stanford, CA 94305-9010, USA, email: khatib@cs.stanford.edu

Professor Frans Groen, Department of Computer Science, Universiteit van Amsterdam, Kruislaan 403, 1098 SJ Amsterdam, The Netherlands, email: groen@science.uva.nl

STAR (Springer Tracts in Advanced Robotics) has been promoted under the auspices of EURON (European Robotics Research Network)

Authors

Dr. Karl Iagnemma
Prof. Dr. Steven Dubowsky
Massachusetts Institute of Technology (MIT)
Department of Mechanical Engineering
77 Massachusetts Avenue
Cambridge, MA 02139-4307
USA

ISSN 1610-7438

ISBN 3-540-21968-4   Springer Berlin Heidelberg New York

Library of Congress Control Number: 2004106986

This work is subject to copyright. All rights are reserved, whether the whole or part of the material is concerned, specifically the rights of translation, reprinting, reuse of illustrations, recitation, broadcasting, reproduction on microfilm or in other ways, and storage in data banks. Duplication of this publication or parts thereof is permitted only under the provisions of the German Copyright Law of September 9, 1965, in its current version, and permission for use must always be obtained from Springer-Verlag. Violations are liable to prosecution under German Copyright Law.

Springer is a part of Springer Science+Business Media

springeronline.com

© Springer-Verlag Berlin Heidelberg 2004
Printed in Germany

The use of general descriptive names, registered names, trademarks, etc. in this publication does not imply, even in the absence of a specific statement, that such names are exempt from the relevant protective laws and regulations and therefore free for general use.

Typesetting: Digital data supplied by authors.
Data-conversion and production: PTP-Berlin Protago-TeX-Production GmbH, Germany
Cover-Design: design & production GmbH, Heidelberg
Printed on acid-free paper   62/3020Yu - 5 4 3 2 1 0

Editorial Advisory Board

EUROPE

Herman Bruyninckx, KU Leuven, Belgium
Raja Chatila, LAAS, France
Henrik Christensen, KTH, Sweden
Paolo Dario, Scuola Superiore Sant'Anna Pisa, Italy
Rüdiger Dillmann, Universität Karlsruhe, Germany

AMERICA

Ken Goldberg, UC Berkeley, USA
John Hollerbach, University of Utah, USA
Lydia Kavraki, Rice University, USA
Tim Salcudean, University of British Columbia, Canada
Sebastian Thrun, Carnegie Mellon University, USA

ASIA/OCEANIA

Peter Corke, CSIRO, Australia
Makoto Kaneko, Hiroshima University, Japan
Sukhan Lee, Sungkyunkwan University, Korea
Yangsheng Xu, Chinese University of Hong Kong, PRC
Shin'ichi Yuta, Tsukuba University, Japan

Foreword

At the dawn of the new millennium, robotics is undergoing a major transformation in scope and dimension. From a largely dominant industrial focus, robotics is rapidly expanding into the challenges of unstructured environments. Interacting with, assisting, serving, and exploring with humans, the emerging robots will increasingly touch people and their lives.

The goal of the *Springer Tracts in Advanced Robotics (STAR)* series is to bring, in a timely fashion, the latest advances and developments in robotics on the basis of their significance and quality. It is our hope that the wider dissemination of research developments will stimulate more exchanges and collaborations among the research community and contribute to further advancement of this rapidly growing field.

The monograph written by Karl Iagnemma and Steve Dubowsky is an evolution of the first Author's Ph.D. dissertation. Mobile robotic systems have lately been receiving a great deal of attention, thanks to their increased use in unstructured environments, such as rugged fields, mines, forests, disaster sites and, last but not least, planetary surfaces after the recent success of the Mars exploration rover missions. This volume addresses several critical problems associated with estimation, motion planning, and control of wheeled mobile robots operating in rough terrain. The unique feature of the work lies in its comprehensive treatment of the problem from the theoretical development of the various schemes to simulation and experimental results for a number of outdoor applications.

The first monograph based on a US doctoral thesis to make the series, this title constitutes a fine addition to STAR!

Naples, Italy *Bruno Siciliano*
March 2004 *STAR Editor*

Preface

New and exciting applications continue to be found for mobile robotic systems. One of the most important trends of the past decade has been the increased use of mobile robots in rough, unstructured terrain, such as underground mines, forests, disaster sites, and planetary surfaces. The extremely successful Sojourner and Mars Exploration Rover missions in 1997 and 2004, respectively, are highly-publicized examples of this trend. Many lesser-known systems have also been developed and successfully deployed.

Unstructured environments are often harsh, dangerous, or inaccessible to humans, and thus motivate the use of robotic systems. For example, planetary surface exploration is hazardous to humans due to high levels of radiation, extreme temperatures, and other environmental factors. Underground mines have traditionally been associated with health hazards caused by poor air quality. These dangers have spurred many robotics researchers to shift their focus from developing robots that operate in indoor, laboratory settings, to those that can successfully overcome the real-world challenges associated with rough terrain operation.

Developing mobile robots for outdoor applications is difficult for several reasons. These operating environments require robots to travel over rugged terrain without becoming entrapped or tipping over. This requires a robot to accurately assess its mobility characteristics over various terrain types and adapt its control strategy accordingly. Outdoor applications also require a robot to depend on simple, on-board sensors to perceive the environment. Such sensors can be noisy, miscalibrated, or unreliable. Finally, outdoor applications often require a robot to possess some degree of autonomy to operate effectively. Specifically, robots often must plan their route through difficult terrain without human supervision.

This monograph addresses several critical problems associated with estimation, motion planning, and control of robotic systems in rough terrain. All of these areas have been studied by various researchers over the past twenty years; however, much of the research has been developed for robots operating in indoor, structured environments. Thus the work in this monograph represents a new view of several "traditional" robotics research areas.

This monograph is composed of five chapters. The first chapter serves as an introduction and overview of the work, and summarizes related research. Chap. 2 addresses the issues of rough terrain modeling and estimation. A novel method for on-line estimation of important terrain physical parameters is presented, as is a novel method for estimating wheel-terrain contact angles. Chap. 3 addresses the problem of rough terrain motion planning by presenting two motion planning algorithms. The goal of the first planning method is to find a safe, direct path from the rover's current position to a distant goal position. The goal of the second planning method is to determine the optimal configuration of an actively articulated suspension rover, to improve tipover stability during travel in rough terrain. Chap. 4 addresses the problem of rough terrain control by presenting a servo-level control method for improved wheel traction or reduced power consumption in rough terrain. Chap. 5 summarizes the contributions of this monograph and presents suggestions for future work. For all of the work, simulation and experimental results are presented for wheeled mobile robots operating in rough, outdoor environments. These results demonstrate the effectiveness of the proposed methods.

The concepts presented in this monograph were initially developed in the first author's 2001 Ph.D. thesis at the Massachusetts Institute of Technology. Much of the work, however, has been updated and expanded for this volume. This work was supported by the NASA Jet Propulsion Laboratory (JPL). The authors would like to acknowledge the assistance and encouragement of Drs. Paul Schenker, Samaad Hayati, and Rich Volpe of JPL.

It is the authors' hope that this monograph will be of interest to robotics researchers and engineers who study and develop robotic systems designed for the outdoor world. As mobile robots become more advanced and their use more widespread, we feel that this subject area can only grow in importance.

Cambridge,   Karl Iagnemma
March 2004

Table of Contents

Chapter 1 Introduction 1
  1.1 Problem Statement and Motivation 1
  1.2 Purpose of This Monograph 3
  1.3 Background and Literature Review 4
    1.3.1 Rough Terrain Modeling and Estimation 4
    1.3.2 Rough Terrain Motion Planning 7
    1.3.3 Rough Terrain Control 11
  1.4 Outline of This Monograph 13
  1.5 Assumptions 13

Chapter 2 Rough Terrain Mobile Robot Modeling and Estimation ... 17
  2.1 Introduction 17
  2.2 Robot Kinematic and Force Analysis 18
    2.2.1 Robot Kinematic Analysis 18
    2.2.2 Robot Force Analysis 21
  2.3 Terrain Characterization and Identification 24
    2.3.1 Equation Simplification 27
    2.3.2 Sensing and Implementation Issues 31
  2.4 Results: Terrain Identification 33
    2.4.1 Simulation Results 33
    2.4.2 Experimental Results 35
  2.5 Wheel-Terrain Contact Angle Estimation 39
    2.5.1 Extended Kalman Filter Implementation 42
  2.6 Results: Wheel-Terrain Contact Angle Identification 44
    2.6.1 Simulation Results 44
    2.6.2 Experimental Results 45
  2.7 Summary and Conclusions 50

Chapter 3 Rough Terrain Motion Planning 51
  3.1 Introduction 51
  3.2 Rough Terrain Motion Planning 52
    3.2.1 Step One: Rapid Path Search 52
    3.2.2 Step Two: Model-Based Evaluation 57
    3.2.3 Uncertainty in Rough Terrain Motion Planning 59

3.2.4. Incorporating Uncertainty in the Rapid Path Search 61
3.2.5. Incorporating Uncertainty in the Model-Based Evaluation.... 62
3.3 Simulation Results—Rough Terrain Planning 65
3.4 Rough Terrain Articulated Suspension Configuration Planning 70
3.4.1 Articulated Suspension Configuration Planning Problem
Description 70
3.4.2 Mobility Analysis 71
3.4.3 Articulated Suspension Configuration Planning
for Enhanced Tipover Stability 72
3.5 Results—Rough Terrain Articulated Suspension Configuration
Planning 75
3.5.1 Simulation Results 75
3.5.2 Experimental Results 76
3.6 Summary and Conclusions 79

Chapter 4 Rough Terrain Control 81
4.1 Introduction 81
4.2 Mobile Robot Rough Terrain Control (RTC) 82
4.3 Wheel-Terrain Contact Force Optimization 84
4.3.1 Optimization Criteria 84
4.3.2 Problem Constraints 86
4.4 Results—Rough Terrain Control 87
4.4.1 Simulation Results 87
4.4.2 Experimental Results 92
4.5 Summary and Conclusions 96

Chapter 5 Conclusions and Suggestions for Future Work 97
5.1 Contributions of This Monograph 97
5.2 Suggestions for Further Work 98

References 101

Index 109

Chapter 1
Introduction

1.1 Problem Statement and Motivation

Mobile robots are increasingly being employed in rough, outdoor terrain for applications such as forestry, mining, search and rescue, and hazardous site inspection (Osborn 1989; Gonthier and Papadopoulos 1998; Cunningham et al. 1999; Mae et al. 2000). These applications often require robots to travel across unprepared, rugged terrain to inspect a location or transport material. In outdoor settings, a robot's mobility is strongly influenced by the terrain geometry and physical properties (Bekker 1956; Wong 1976). For example, a robot traversing loose, sloped sand might experience substantial wheel slippage and poor mobility, whereas a robot traversing flat, firm clay might experience excellent mobility. Outdoor operation also often requires robots to rely on on-board sensors (such as rangefinders and inertial measurement units) for navigation and control. These sensors generally contain significant uncertainty and error in their measurements. Finally, outdoor applications often require robots to operate autonomously, which requires real-time decision making that is constrained by limited on-board computational resources. The combined effects of rough, varying terrain conditions, sensor measurement uncertainty and error, and limited computation make rough terrain motion planning and control challenging problems.

An important application for mobile robots is planetary exploration (Hayati et al. 1996; Schenker et al. 1997; Weisbin et al. 1999). In 1997 and 2004, small wheeled robots ("rovers") landed on the surface of Mars to conduct scientific experiments focused on understanding the planet's climate history, surface geology, and potential for past or present life. These NASA / JPL rovers were expected to negotiate loose, sloped, and rock-filled terrain to access scientifically interesting sites. The 1997 Sojourner rover limited its traverses to relatively short distances (i.e. a few meters) through flat, rocky terrain under close human supervision (see Fig. 1.1). The 2004 Mars Exploration Rovers are, as this monograph goes to press, performing longer-distance traverses through loose, sloped terrain, with some autonomous operation. In general, these rovers were designed to negotiate moderately rough terrain under close guidance by human operators.

Goals for future Mars exploration missions are highly ambitious from a robotic mobility perspective (Volpe 2003). Science objectives are expected to focus on exploration of geographical features that were possibly shaped by water (Carr 1996). This includes terrain areas such as gullies, outflow channels, and hydrothermal regions. These areas are often rough and sloped, with dense rock distributions and loose drift material that presents severe challenges to the mobility capabilities of current rovers (Bernard and Golombek 2001). In such environments, a rover could experience severe wheel slippage or sinkage, or get "hung up" on a large rock, which could lead to entrapment.

Fig. 1.1. Sojourner rover on the surface of Mars, 1997 (NASA Jet Propulsion Laboratory)

To accomplish these challenging objectives, future robot designs may evolve from traditional "fixed configuration" designs to those with actively articulated suspensions (Schenker et al. 2000). Robots with articulated suspensions can improve rough terrain mobility by modifying their suspension configuration and thus repositioning their center of mass. This allows access to challenging terrain regions with a higher degree of stability than is otherwise possible. Although these advanced suspension designs will yield improved rough terrain mobility, the dominant factors in robot mobility remain the terrain geometry and physical properties. Thus advanced robot designs must be accompanied by motion planning and control algorithms that specifically consider robot-terrain interaction.

Most current mobile robot motion planning and control algorithms are not well suited to rough terrain environments, since they generally do not consider the physical interaction of the robot and terrain. This is likely due to the fact that terrain parameters are difficult to directly measure. Also, current planning and control algorithms often assume that the robot has perfect sensory knowledge of the environment, which is never the case for these applications. Failure to account for robot-terrain interaction and sensor uncertainty can lead to system failure through loss of mobility or obstacle entrapment. Alternatively, the robot could exhibit unnecessarily conservative behavior. This could limit the ability of robots to perform valuable tasks, such as reaching science targets that may be located in challenging terrain.

In summary, mobile robots operating in rough terrain must understand the physical properties of the terrain they are traversing, and account for uncertainty inherent in their sensing systems. These robots may have "traditional" designs or actively articulated suspensions. They must be able to accomplish planned tasks with some degree of autonomy, while ensuring robot safety.

1.2 Purpose of This Monograph

The purpose of this monograph is to present methods for improving mobile robot mobility in high-risk, rough terrain environments, through the development of estimation, motion planning, and control algorithms that rely on physical models of the robot and terrain. Rough terrain is defined here as unstructured terrain that includes features that could cause robot entrapment or loss of stability. This includes regions of loose soil, dense rocks or boulders, steeply sloped terrain, or a combination of all of these features. Rough terrain could occur either outdoors, such as on a planetary surface, or indoors, such as in a collapsed building. This monograph presents methods and algorithms related to three basic problems in rough terrain robot mobility: estimation, motion planning, and control. The primary intended application area for this work is planetary exploration. However, the methods and algorithms presented here can be applied to a wide range of rough terrain situations.

Estimation algorithms are presented that focus on robot-terrain interaction. Specifically, methods are presented for estimating terrain cohesion and internal friction angle (two important physical parameters related to terrain traversability), and wheel-terrain contact angles. A model of robot-terrain interaction mechanics is also presented, which unites the estimated terrain quantities with "classical" robot models. The purpose of this work is to allow a robot to accurately and autonomously assess whether or not a proposed terrain region can be safely traversed. The estimated quantities will also be used as inputs to motion planning and control algorithms.

Algorithms are presented for two distinct aspects of the motion planning problem. Both methods rely on models and estimation methods developed earlier in the monograph. First, a path planning algorithm that determines a short, traversable path through rough terrain is presented. Second, an algorithm for planning the suspension configuration of actively articulated suspension robots is presented. This algorithm maximizes the stability of the vehicle and allows it to safely access highly sloped and rough terrain.

Finally, a rough terrain control algorithm is presented. The algorithm uses knowledge of terrain parameters and wheel-terrain contact angles to minimize wheel slip and improve traction. The purpose of this algorithm is to allow a robot to safely and efficiently traverse very rough terrain.

1.3 Background and Literature Review

In this section a summary of literature related to the work in this monograph is presented. This review is divided into sections focusing on modeling and estimation, motion planning (including path planning and articulated suspension configuration planning), and control.

1.3.1 Rough Terrain Modeling and Estimation

Modeling of mobile robots has been studied by many researchers in various contexts. Kinematic and dynamic analysis of robots on flat, structured terrain is a well-understood problem (Muir 1987; Sarkar et al. 1994; Laumond 1998; Mutambara and Durrant-Whyte, 2000). Kinematic analysis of robotic systems in rough terrain has been addressed more recently in (Milesi-Beller et al. 1993; Sreenivasan and Nanua 1996; Sreenivasan and Waldron 1996; Gajjar and Johnson, 2002). Specific kinematic analyses of six-wheeled robots with rocker-bogie suspensions (such as the JPL Sojourner rover) have been presented in (Chottiner 1992; Linderman and Eisen 1992; Hacot 1998; Tarokh et al. 1999). Force analyses of mobile robots has also been performed. The mobile robot force analysis problem is similar to the force distribution problem in closed kinematic chains and walking machines, which has been studied in (Kumar and Gardner 1990; Kumar and Waldron 1990). Active coordination of forces in wheeled systems was first proposed in (Kumar and Waldron 1989), and was later addressed in (Sreenivasan 1994; Sreenivasan and Nanua 1996). In general, there is a significant body of literature related to kinematic and dynamic modeling of mobile robots in both flat and uneven terrain. This monograph does not focus on kinematic and dynamic modeling per se, but instead addresses important related problems.

An important and often neglected aspect of robot system modeling is wheel-terrain interaction phenomena. Wheel-terrain interaction plays a critical role in rough terrain mobility, since traction characteristics are governed by terrain physical properties (Bekker 1956, 1969; Wong 1976). For example, a robot traveling through loose sand has very different mobility characteristics than one moving across firm clay. Pioneering research in this area was performed by Bekker (Bekker 1956; Bekker 1969). This work focused on developing fundamental models of wheel- (and track-) terrain interaction. To apply these models in the field, equipment and techniques were developed for estimating important model parameters.

Numerous researchers have studied terrain parameter estimation. Generally, these methods involve off-line estimation using dedicated testing equipment (Nohse et al. 1991; Shmulevich et al. 1996). Parameter estimation for a legged system has been studied in (Caurin and Tschichold-Gurman 1994). This approach relies on feedback from an embedded multi-axis force sensor in the robot's leg, and is not applicable to wheeled systems. Terrain parameter estimation for tracked vehicles has been proposed in (Le et al. 1997). This approach assumes a simplified "force coefficient" model of track-terrain interaction that is not valid in deformable terrain. Other researchers have attempted to circumvent explicit parameter estimation by using simplified models to estimate wheel slip or drawbar pull (Lindgren et al. 2002; Yoshida et al. 2002; Yoshida et al. 2003). Control algorithms designed to minimize slip or maximize drawbar pull can then be employed. However, estimation of terrain parameters allows a richer physical description of wheel-terrain interaction than does estimation of slip alone.

In addition to its utility from a mobility perspective, estimation of terrain parameters is an important scientific goal of planetary surface exploration missions (Moore et al. 1977, 1999; Matijevic et al. 1997). Parameter estimation of Martian soil has been performed by the Viking landers and the Sojourner rover, and is currently being performed by the two Mars Exploration Rovers, Spirit and Opportunity (see Fig. 1.2). The Viking landers used manipulator arms to conduct soil trenching experiments (Moore et al. 1977). The Sojourner rover used its wheels as trenching devices to identify soil cohesion and internal friction angle, among other parameters (Matijevic et al. 1997). Both missions used visual cues and off-line (i.e. Earth-based) analysis techniques to compute soil parameters. Earth-based analysis requires lengthy communication time delays that reduce rover efficiency and limit its autonomy.

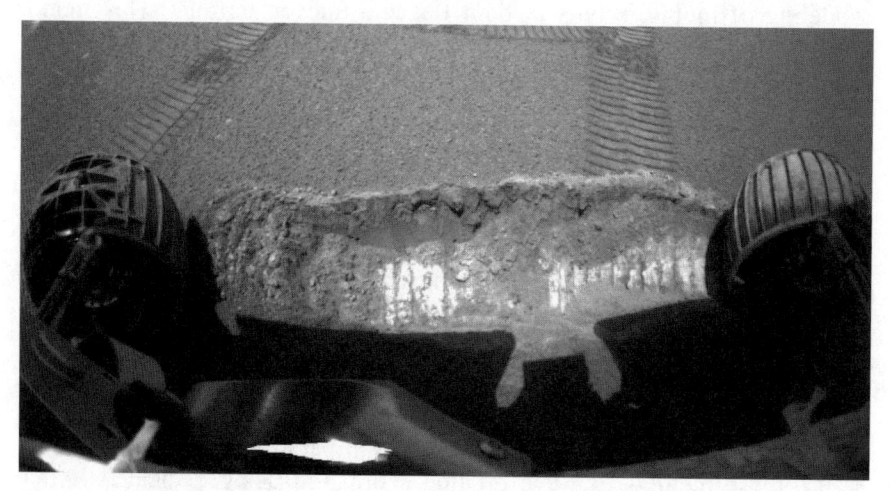

Fig. 1.2. Wheel trench dug by Mars Exploration Rover Spirit for terrain parameter estimation experiment (NASA Jet Propulsion Laboratory)

It is important to estimate terrain physical parameters on-line, since on-line estimation would allow a robot to predict its ability to safely traverse changing terrain (Iagnemma et al. 2002). It would also allow a robot to adapt its control and planning strategies to variable terrain conditions, to maximize wheel traction or minimize power consumption (Iagnemma and Dubowsky 2000b). In this monograph, a method for on-line parameter estimation of terrain the robot is currently traversing is presented. This allows accurate assessment of traversability. Additionally, the parameters can be used as inputs to terrain models which form a basis for enhanced rough terrain motion planning and control algorithms.

Another important element of a robot system model is wheel-terrain contact angles, since they greatly influence robot traversability properties (see Fig. 1.3). For example, a robot traversing flat, even terrain has very different mobility characteristics than one traversing steep, uneven terrain.

This difference can in part be analyzed and predicted by sensing or estimating the wheel-terrain contact angles. Previous researchers have proposed installing multi-axis force sensors at each wheel hub to measure the contact force direction (Sreenivasan and Wilcox 1994). Wheel-terrain contact angles could be inferred from the contact force direction. However, installing multi-axis force sensors at each wheel is costly and mechanically complex. A method for contact angle estimation has been proposed that is based on knowledge of the terrain map (Balaram 2000). However, for deformable or vegetation-covered terrain the terrain map (as measured by a laser or stereo-based range sensor) may not correspond to the geometry experienced by the robot. This method is also computationally intensive. In this monograph a method is presented for wheel-terrain contact angle estimation that utilizes simple on-board sensors and is computationally efficient (Iagnemma and Dubowsky 2000a).

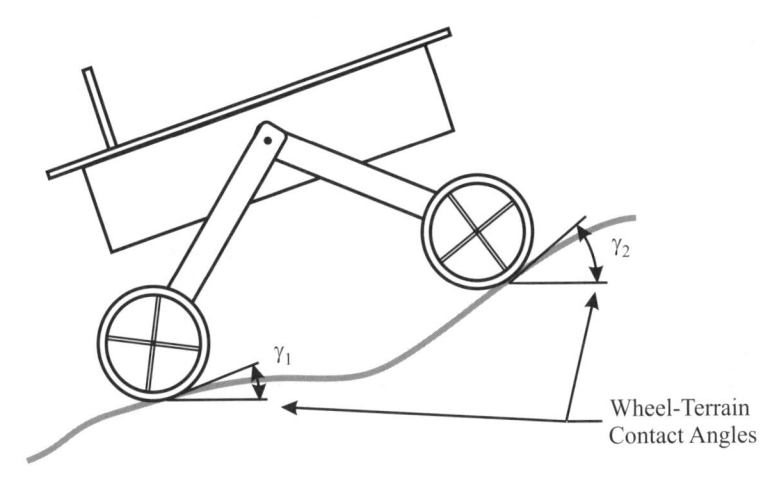

Fig. 1.3. Wheel-terrain contact angles

1.3.2 Rough Terrain Motion Planning

Path Planning

Many rough terrain applications require mobile robots to autonomously determine a safe route to a distant position, and safely traverse difficult terrain to reach that position. This might be a material loading station, a location in a collapsed building, or a science target on a planetary surface. The problem of finding a route from the robot's current position to a goal position is referred to as the motion planning problem. Numerous planning methods have been proposed over the past twenty years, using techniques such as quadtrees, graph-search methods, potential fields, and fuzzy logic (see, for example, Khatib 1986; Warren 1993; Stentz 1994; Haddad et al. 1998; Yahja et al. 1998; Lee and Wu 2003). These methods generally consider the case of a two or three degree-of-freedom robot moving in a planar environment. A survey of many "traditional" planning methods can be found in (Latombe 1991).

Many traditional motion planning methods cannot be successfully applied in rough terrain environments, since they often assume perfect knowledge of the environment, ignore vehicle and terrain mechanics, and represent obstacles and free space in a binary format (Latombe 1991). Additionally, many traditional planning methods are computationally expensive. These factors are critical to rough terrain motion planning for several reasons. First, in rough terrain the concept of an obstacle is not clearly defined, as it depends on an understanding of the terrain and the mobility characteristics of the robot. Thus there exists a continuous (rather than binary) gradation between easily traversed terrain and impassable obstacles. If a simplistic obstacle definition is employed a mobile robot will likely either act with unnecessary caution or with excessive risk. Second, terrain data cannot be assumed to be perfectly known, due to errors in range sensing techniques (Hebert and Krotkov 1992; Matthies and Grandjean 1994). These errors are caused by miscalibration and inherent error in the sensing device. Third, the planned path might not be accurately followed by the robot due to path-following errors (Volpe 1999). Finally, autonomous mobile robots will generally have limited computational resources to devote to path planning.

Recently, researchers have begun addressing the rough terrain planning problem. First works were dedicated to computing dynamic, time-optimal paths through undulating terrain (Shiller and Chen 1990; Shiller 2000). Other researchers have utilized simplified dynamic vehicle models to ensure that proposed paths do not cause vehicle tipover (Olin and Tseng 1991; Kelly and Stentz 1998). Employing a kinematic model to evaluate traversability at a large number of points in the configuration space of the robot's position and heading has been proposed in (Siméon and Dacre-Wright 1993; Cherif and Laugier 1994; Farritor et al. 1998(a); Farritor 1998(b); Cherif 1999; Bonnafous et al. 2001; Kubota et al. 2001; Spero and Jarvis 2002). These methods generally employ search methods to find a suitable path. The methods differ largely in the type of search technique they employ. Proposed search techniques include the A^* and D^* algorithms, genetic algorithms, and rapidly-exploring random trees (RRTs).

Model-based slip-free motion planning for an articulated vehicle has been proposed (Choi and Sreenivasan 1998). All of these methods recognize the importance of model-based analysis to ensure path traversability. However, they generally utilize simplified terrain models and do not consider uncertainty.

Another class of algorithms is based largely on finding the smoothest path through a given terrain region. The smoothest path is assumed to be the most traversable. One approach uses fractals to model terrain, and searches for a path with a consistently low fractal dimension (Pai and Reissel 1998). Another method models obstacles with potential fields, and searches for a path with low potential (Chanclou and Luciani 1996). A fuzzy logic-based method has been proposed that uses gross knowledge of terrain slope and roughness to avoid hazardous regions (Seraji 1999; Seraji and Howard 2002). A sensor-based method has been implemented that classifies obstacles in a binary manner to determine an obstacle-free path (Laubach et al. 1998; Laubach and Burdick 1999). In general, these methods do not consider vehicle mechanics, or allow for uncertainty. They attempt to avoid highly rough terrain, and implicitly assume that the planned path is free of hazards. (It should be noted that flat terrain regions are not always traversable, since "non-geometric obstacles" such as loose soil may be untraversable. Reliable detection of these non-geometric obstacles is an open research problem.) Thus, the above methods may be effective in flat terrain with "discrete" obstacles, such as large boulders, but may not be well suited to rough, uneven terrain.

In summary, with the exception of (Ben Amar and Bidaud 1995), most proposed planning methods do not employ a realistic terrain model. This is critical to an accurate assessment of terrain traversability. Additionally, with the exception of (Gifford and Murphy 1996; Hait and Siméon 1996), most proposed planning methods do not consider uncertainty in the terrain data or robot path-following accuracy. In rough terrain, failure to account for uncertainty can cause a robot to travel into unknown terrain regions, with potentially hazardous consequences. In this monograph a motion planning method is presented that utilizes model-based analysis of robot-terrain interaction, and considers terrain data and path-following uncertainty (Iagnemma et al. 1999a). It is also computationally efficient enough for on-board implementation.

Articulated Suspension Configuration Planning

Motion planning algorithms generate a path through a terrain region to a goal location. The robot must then safely traverse the path without tipping over, getting "hung up" on an obstacle, or becoming immobilized in loose soil. Robots with articulated suspensions can improve rough terrain mobility by modifying their suspension configuration and thus repositioning their center of mass. This allows them safe access to more difficult terrain. Articulated suspensions can either be passive or active. The most notable example of a passively articulated suspension designed for rough terrain mobility is the rocker-bogie suspension employed on the JPL family of planetary exploration rovers (Bickler 1992). Other researchers have developed similar systems (Kawakami et al. 2002; Kimura and Hirose 2002; Siegwart et al. 2002).

Actively articulated suspensions are generally more complex than passively articulated suspensions, but can yield greatly improved rough terrain mobility. One example of an actively articulated suspension robot is the Jet Propulsion Laboratory's Sample Return Rover (SRR), shown in Fig. 1.4. The SRR can actively modify its two shoulder joints to change its center of mass location and enhance rough terrain mobility. For example, when traversing a side slope the SRR can adjust angles θ_1 and θ_2 to improve stability, as illustrated in Fig. 1.5. It can also reposition its center of mass by moving its manipulator.

Previous researchers have suggested the use of articulated suspension systems to enhance rough terrain mobility. First works focused on the four-wheeled JPL GOFOR (Sreenivasan and Wilcox 1994). This work relied on analysis of the GOFOR's particular configuration, and is therefore difficult to apply to general mobile robots. Additionally, it considered only planar vehicle motion, and was not demonstrated experimentally. Other work studied articulated suspension control for the Wheeled Actively Articulated Vehicle (WAAV) to allow difficult mobility maneuvers (Sreenivasan and Waldron 1996). Solutions for the WAAV's specific kinematic structure were presented, and studied in simulation. Other researchers have developed genetic algorithms for repositioning a vehicle's manipulator to modify its center of mass location and aid mobility (Farritor et al. 1998b). This approach was computationally intensive, and thus not suited for on-line operation. None of the previously proposed methods have been demonstrated on an experimental robot system in rough terrain.

In this monograph an efficient method for planning the configuration of an actively articulated suspension is presented and applied experimentally to the JPL Sample Return Rover (SRR) (Huntsberger et al. 1999; Iagnemma et al. 2000(c); Iagnemma et al. 2003). Simulation and experimental results for the SRR under field conditions show that articulated suspension control can greatly improve rover stability in rough terrain.

Fig. 1.4. Jet Propulsion Laboratory Sample Return Rover (SRR), shown with actively articulated suspension at two different positions (NASA Jet Propulsion Laboratory)

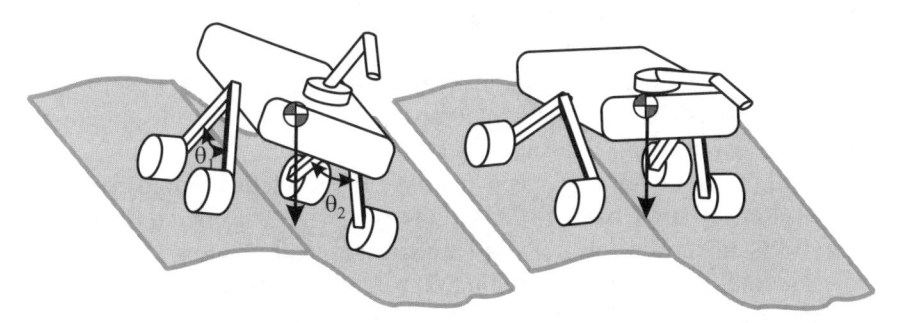

Fig. 1.5. An actively articulated suspension robot improving rough terrain tipover stability by adjusting joint angles θ_1 and θ_2

1.3.3 Rough Terrain Control

In rough terrain, it is critical for mobile robots to maintain adequate wheel traction. Excessive wheel slip could cause a robot to roll over or deviate from its intended route. Also, if excessive force is applied to a terrain region, the soil may fail and induce excessive wheel sinkage, which may lead to entrapment. Substantial work has been performed on traction control of passenger vehicles operating on flat roads (Mohan and Williams

1995; Kawabe et al. 1997; Van Zanten et al. 1997; Van Zanten et al. 1998). These approaches rely on mechanical torque distribution systems, such as differentials, which mobile robots generally are not equipped with. Fuzzy logic wheel-slip control has been proposed for passenger vehicles on paved roads (Mauer 1995; Cheok et al. 1997). These methods assume that the vehicle forward velocity is known, which allows computation of wheel slip. The wheel slip is then used as a control variable. However, forward velocity is difficult to measure or estimate for slow-moving mobile robots, since at low speeds measurements from inertial measurement units are susceptible to noise. Recent state estimation techniques based on visual feedback have shown promise for estimating robot forward velocity at low speeds (Olson et al. 2001).

Researchers have proposed a variable-structure control approach for traction control of passenger vehicles on paved roads that does not utilize a mechanical differential (Tan and Chin 1991; Tan and Chin 1992; Lee and Tomizuka 1996). However, these approaches assume a form of the traction-slip ratio relationship that is valid only for deformable tires on hard terrain. In off-road situations the wheel is often rigid while the terrain is deformable. The resulting mechanics are substantially different from the deformable tire/rigid terrain case. Additionally, this work is only partially applicable to rough terrain situations, since in rough terrain wheel slip is primarily caused by kinematic incompatibility or loose soil conditions, rather than "breakaway" wheel acceleration.

Traction control for low-speed mobile robots on flat terrain has been studied (Reister and Unseren 1993). Later work has considered the important effects of terrain unevenness (Sreenivasan and Wilcox 1994). This work assumes knowledge of terrain geometry and soil characteristics, and has not been validated experimentally. Terrain geometry can be measured with standard range sensors; however, as noted above, the measured terrain geometry may not correspond to the geometry experienced by the robot for deformable or vegetation-covered terrain. Additionally, in most applications the soil characteristics are unknown. A fuzzy-logic traction control algorithm for a rocker-bogie rover that does not assume knowledge of terrain geometry has been developed (Hacot 1998). This approach is based on heuristic rules related to vehicle mechanics, and again assumes that the wheel slip ratio is measurable, which is generally not true for slow-moving robots. Other researchers have proposed a method for minimizing wheel slip by synchronizing robot wheel velocities in uneven terrain (Peynot and Lacroix 2003). This work relies on assumptions regarding the slip state of a wheel.

In this monograph a rough terrain control method is presented that utilizes simple sensory inputs to optimize wheel torque for maximum traction or minimum power consumption, depending on the local terrain difficulty (Iagnemma and Dubowsky 2000b). It does not rely on mechanical torque distribution systems or measurement of wheel slip. Simulation and experimental results demonstrate the effectiveness of the method.

1.4 Outline of This Monograph

This monograph is composed of five chapters. This chapter serves as an introduction and overview of the work, and summarizes related research.

Chap. 2 addresses rough terrain modeling and estimation by presenting models for mobile robot kinematic analysis, force analysis, and wheel-terrain interaction. A method for on-line estimation of important terrain physical parameters is presented. A method for estimating wheel-terrain contact angles from on-board sensors is also presented. Simulation and experimental results are presented for a six-wheeled rover in rough, sandy terrain.

Chap. 3 addresses rough terrain motion planning by presenting two motion planning algorithms. The goal of the first planning method is to find a safe, direct path from the rover's current position to a goal position several rover lengths distant. The goal of the second planning method is to determine the optimal configuration of an actively articulated suspension rover, to improve tipover stability during travel in rough terrain. Simulation and experimental results are presented for the JPL SRR operating in outdoor terrain.

Chap. 4 addresses rough terrain control by presenting a servo-level control method for improved wheel traction or reduced power consumption in rough terrain. Simulation and experimental results are presented for a six-wheeled rover in rough, sandy terrain.

Chap. 5 summarizes the contributions of this monograph and presents suggestions for future work.

1.5 Assumptions

There are three primary assumptions made in the work presented in this monograph. The first is that the robotic systems under consideration move slowly and thus forces arising from dynamic effects are small. As a result, the methods and algorithms presented here rely on kinematic (rather than dynamic) analysis. This assumption was motivated by the observation that a broad range of rough terrain systems, including planetary exploration rovers, move at speeds on the order of 10 cm/sec. At these speeds, forces due to robot dynamics are negligible. Recent research into higher-speed systems (for applications such as military reconnaissance) has considered systems moving at speeds on the order of 10 m/s or higher. At these speeds on uneven terrain, robot dynamics play a significant role and cannot be ignored. The extension of some of the methods described in this monograph to high-speed robotic systems is an area of current research.

The second assumption is that the robot wheels are stiff relative to the terrain. This analysis is one of four possible wheel-terrain interaction cases. Other cases are a rigid wheel traveling over rigid terrain, a deformable wheel traveling over rigid terrain, and a deformable wheel traveling over deformable terrain, as shown in Fig. 1.6.

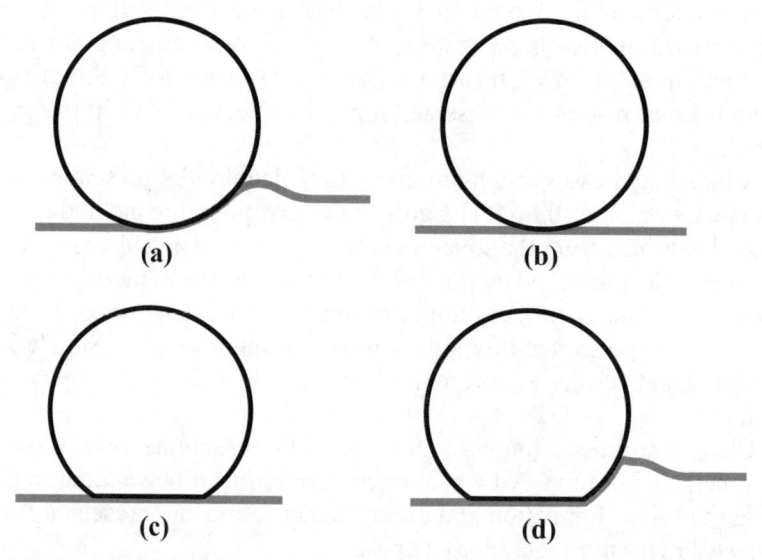

Fig. 1.6. Four cases of wheel-terrain interaction mechanics: *(a)* rigid wheel traveling over deformable terrain, *(b)* rigid wheel traveling over rigid terrain, *(c)* deformable wheel traveling over rigid terrain, and *(d)* deformable wheel traveling over deformable terrain

It is important to distinguish between these cases, as fundamental wheel-terrain mechanics vary depending on the interaction conditions (Bekker 1969; Plackett 1985; Wong 1976). Here the case of a rigid wheel in deformable terrain is examined, as this is the condition for many rough terrain systems, including planetary exploration rovers. Note that this case is common in terrestrial vehicles, since a pneumatic tire can be considered rigid if its inflation pressure is high compared to the terrain stiffness (Bekker 1969). Proper analysis of a wheel-terrain system depends on the relative stiffness of both the wheel and terrain.

The third assumption used in this monograph is that wheel-terrain contact can be modeled as occurring at a single point and that a wheel-terrain contact angle can be defined by the location of that point on the wheel rim. This is valid for the assumed case of a rigid wheel traveling over deformable terrain, since an "effective" wheel-terrain contact angle (and therefore point) can be defined as the angular direction of travel imposed on the wheel by the terrain during motion, as illustrated in Fig. 1.7. This assumption is also clearly valid for the case of a rigid wheel traveling over rigid terrain, described above.

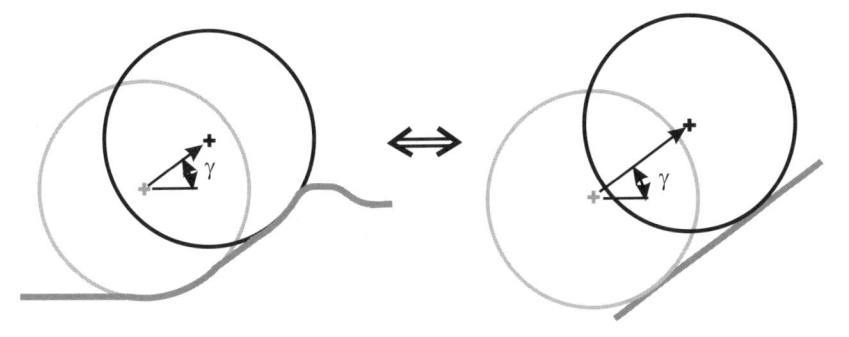

Fig. 1.7. Effective wheel-terrain contact angle for rigid wheel on deformable terrain, and an analogous case for a rigid wheel on rigid terrain

Chapter 2
Rough Terrain Mobile Robot Modeling and Estimation

2.1 Introduction

Kinematic and dynamic modeling are fundamental aspects of mobile robot analysis, motion planning, and control. Traditionally, for robots operating on level, smooth surfaces (e.g. laboratory, factory, or hospital floors), kinematic and dynamic models rely on knowledge of robot design parameters such as wheel diameter and spacing, inertial parameters, motor characteristics, etc. This information can then be used to develop models that yield insights about robot mobility. Such models can also be used as a basis for motion planning and control algorithms.

In rough terrain, surface characteristics have a significant impact on robot performance. This includes both surface traction characteristics and surface unevenness properties. In applications such as planetary exploration, it is important that motion planning and control algorithms be robust and flexible enough to operate effectively under various terrain conditions. Thus to optimize system performance, some knowledge of the surface characteristics must be considered in the robot model. This represents a fundamental difference between "traditional" robot analysis and rough terrain robotic analysis.

This chapter presents a description of the rough terrain kinematic and force analysis problems, which are integral to many aspects of motion planning and control. Kinematic and force analyses are shown to rely on an understanding of robot wheel-terrain interaction. Specifically, it is shown in Sect. 2.2 that estimating terrain physical parameters and wheel-terrain contact angles are important for accurate kinematic and force analysis. A method for on-line estimation of terrain physical parameters is presented in Sect. 2.3. Simulation and experimental results for terrain parameter estimation are presented in Sect. 2.4. A method for on-line estimation of wheel-terrain contact angles is presented in Sect. 2.5. Simulation and experimental results for contact angle estimation are presented in Sect. 2.6. Sect. 2.7 is a chapter summary and presents conclusions drawn from the work.

It is shown that terrain physical parameters and wheel-terrain contact angles can be accurately estimated using on-board sensors with limited computational effort, despite the presence of noise. This leads to an improved understanding of mobile robot performance in rough terrain.

2.2 Robot Kinematic and Force Analysis

A fundamental aspect of mobile robot analysis is kinematic and dynamic modeling. Numerous researchers have addressed the problem of modeling mobile robots on level, smooth surfaces, as described in Sect. 1.3.1. Here the problems of rough terrain kinematic and force analysis are presented, for the purpose of illustrating challenges relating to rough terrain operation. Note that in this monograph, the general problem of mobile robot modeling is not addressed. Only issues related specifically to rough terrain operation are considered.

2.2.1 Robot Kinematic Analysis

The purpose of kinematic analysis is to determine a robot's kinematic configuration on an arbitrary terrain profile. This is useful for motion planning algorithms that attempt to determine if a robot can physically conform to a given terrain region without violating joint limit or interference constraints. Such a planning algorithm will be presented in Sect. 3.2.

In this work the inverse kinematics problem is of primary interest, and can be informally stated as follows: *Given an elevation map of the terrain and the position of a point on the robot body, compute the orientation (Θ, Φ, Ψ) of the robot body and the configuration $(\theta_1, \ldots, \theta_n)$ of the robot suspension.* The quantities (Θ, Φ, Ψ) are the body roll, pitch, and yaw, respectively, and $(\theta_1, \ldots, \theta_n)$ are variables related to articulation or suspension joints the robot may possess. An illustration of the inverse kinematics problem is shown in Fig. 2.1.

An example of particular interest is the kinematic analysis of a robot with a rocker-bogie suspension, as shown in Fig. 2.2 (Tarokh et al. 1992; Bickler 1992). Similar kinematic configurations have been successfully employed by the Mars Sojourner and Mars Exploration rovers. Such suspensions can be designed to equalize load among wheels during traversal of uneven terrain, thereby maximizing allowable tractive force.

2.2 Robot Kinematic and Force Analysis

Ten parameters are required to define the configuration of the robot in Fig. 2.2:

- The position of a point on the robot body $\mathbf{p}_c = [p_x\ p_y\ p_z]^T$ expressed in an inertial frame $\{XYZ\}$. Here we choose without loss of generality that $\mathbf{p}_c$ is coincident with the robot center of mass;
- The orientation of the robot body (Θ, Φ, Ψ) expressed in $\{XYZ\}$;
- The joint angle values of the rocker-bogie mechanism $(\theta_{1r}, \theta_{2r}, \theta_{1l}, \theta_{2l})$.

The inverse kinematics problem can be stated as follows: *Given a pointwise terrain elevation map $z(x,y)$, the position of the robot center of mass $\mathbf{p}_c$, and the robot heading Ψ, compute the orientation (Θ, Φ) and the joint angle values $(\theta_{1r}, \theta_{2r}, \theta_{1l}, \theta_{2l})$.* Here an Euler angle parameterization is employed in the order Ψ, then Θ, then Φ. Note that in real-world applications, the terrain elevation map (or a sparse estimate thereof) is often measured by on-board rangefinder systems or known *a priori* from satellite or survey measurements.

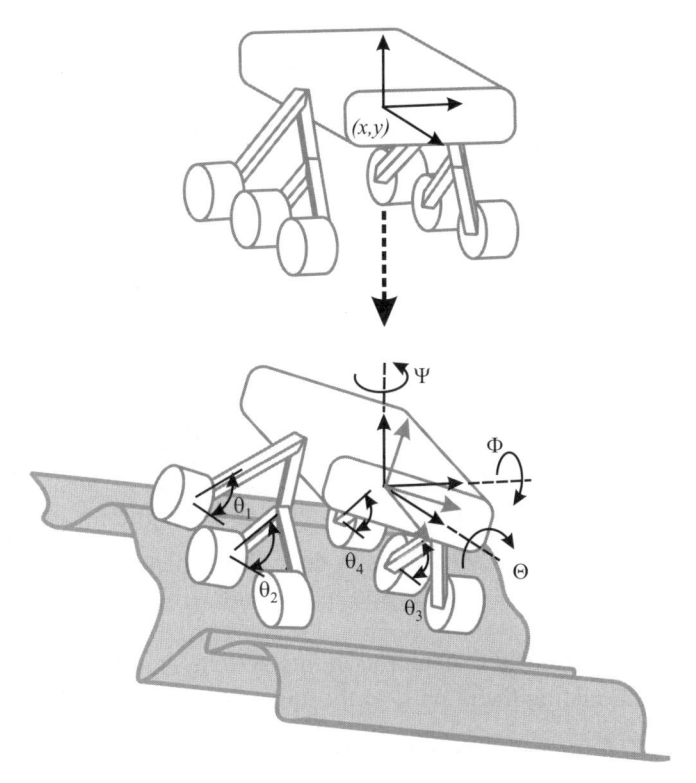

Fig. 2.1. Illustration of robot inverse kinematics problem

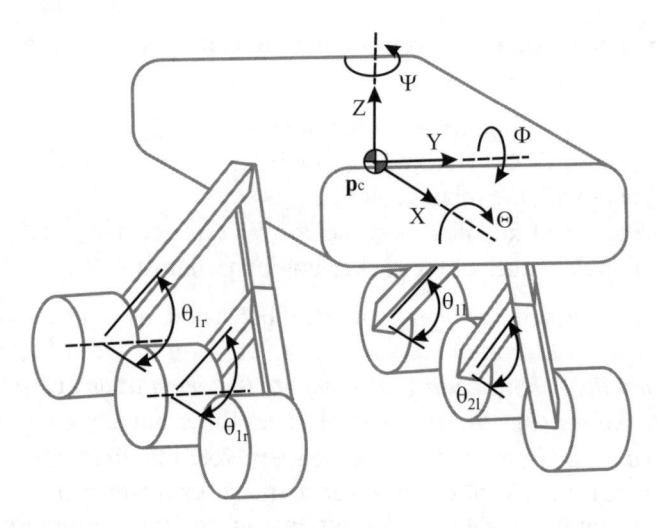

Fig. 2.2. Kinematic description of a six-wheeled robot with a rocker-bogie suspension

For a vehicle with m unique wheel-terrain contact points, at least $m-1$ kinematic loop closure equations can be written (Eckhardt 1989). For the case of robot shown in Fig. 2.2, these loop closure equations can be written as:

$$z_{rr} = z_{lr} + l_1 \cos\Theta(\sin\theta_{1r} - \sin\theta_{1l}) + w\sin\Theta, \tag{2.1}$$

$$z_{rr} = z_{lm} + \cos\Theta(l_1 \sin\theta_{1r} - l_2 \cos\theta_{1l} - l_3 \sin\theta_{2l}) + w\sin\Theta, \tag{2.2}$$

$$z_{rr} = z_{lf} + \cos\Theta(l_1 \sin\theta_{1r} - l_2 \cos\theta_{1l} - l_4 \cos\theta_{2l}) + w\sin\Theta, \tag{2.3}$$

$$z_{rr} = z_{rm} + \cos\Theta(l_1 \sin\theta_{1r} - l_2 \cos\theta_{1r} - l_3 \sin\theta_{2r}), \tag{2.4}$$

$$z_{rr} = z_{rf} + \cos\Theta(l_1 \sin\theta_{1r} - l_2 \cos\theta_{1r} - l_4 \cos\theta_{2r}), \tag{2.5}$$

where z_{ij}, $i = \{r,l\}, j = \{r,m,f\}$ refers to the z component of $\mathbf{p}_{ij}$, with index i referring to the right or left side, and index j referring to the rear, middle, or front wheel.

Due to the mechanical differential in this system, an additional equation can be written relating the pitch Φ to the angles θ_{lr} and θ_{ll}:

$$\Phi = \frac{\theta_{lr} + \theta_{ll}}{2} + \theta'_{lr} \tag{2.6}$$

where θ'_{lr} is the value of θ_{lr} when the robot is on flat terrain. Thus, six unique kinematic equations can be written to describe the robot in Fig. 2.2.

The position $\mathbf{p}_c$ and heading Ψ are taken as inputs since the goal of kinematic analysis is to predict kinematic validity and static stability at a given point in the terrain. Thus, the robot will be "virtually placed" at a point in the terrain map, and kinematic analysis will be performed to determine if the point is kinematically valid. Knowledge of position and heading reduces the number of unknown parameters to six, which can be determined by solving the system of Eqs. (2.1-2.6).

Numerical techniques such as Newton's method and steepest descent can be applied to this problem, although convergence is not guaranteed since the terrain elevation map is generally not represented by a continuously differentiable function. An efficient solution method for robot inverse kinematics problems has been presented in (Hacot 1998). Here we simply point out that wheel-terrain contact on uneven terrain can occur at various locations along the wheel rim, and this location influences the solution to the inverse kinematics problem. Also, robot wheels may experience varying amounts of sinkage, which also affects the inverse kinematics solution.

2.2.2 Robot Force Analysis

The purpose of force analysis is to determine if a robot can exert enough thrust on an arbitrary terrain profile to produce motion in a given direction without violating motor torque saturation or terrain traction constraints. The force analysis problem can be informally stated as follows: *Given the robot configuration and wheel-terrain contact angles, determine if a set of wheel-terrain contact forces exist that can provide a body force vector in the direction of desired motion.*

A six-wheeled mobile robot with a rocker-bogie suspension on uneven terrain is shown in Fig. 2.3. The vectors $\mathbf{f}_i = [f_i^x \; f_i^y \; f_i^z]^T$, $i = \{1,...,6\}$, represent wheel-terrain interaction forces and are expressed in the inertial frame *{XYZ}*. The vectors $\mathbf{P}_i = [p_i^x \; p_i^y \; p_i^z]^T$, $i = \{1,...,6\}$, are directed from the wheel-terrain contact points to the robot center of mass $\mathbf{p}_c$ and are also expressed in the inertial frame *{XYZ}*. The vector $\mathbf{f}_s = [F_s^x \; F_s^y \; F_s^z \; M_s^x \; M_s^y \; M_s^z]^T$ at the center of mass represents the summed effects of gravitational forces, inertial forces, manipulation forces, and forces due to interaction with the environment or other robots. Link, body, and wheel masses are lumped at the center of mass which is assumed to be fixed.

Note that $\mathbf{f}_s$ possesses a user-specified component in the direction of desired motion. Thus, if a set of wheel-terrain interaction force vectors $\mathbf{f}_i$ can

22   Chapter 2   Rough Terrain Mobile Robot Modeling and Estimation

be found can provide a small positive body force vector $\mathbf{f}_s$, the robot can move in the direction of desired motion. This assumes that motion resistance due to terrain effects (such as soil compaction) is small. In cases where motion resistance is expected to be large, $\mathbf{f}_s$ can be chosen to be larger than the estimated net resistance force.

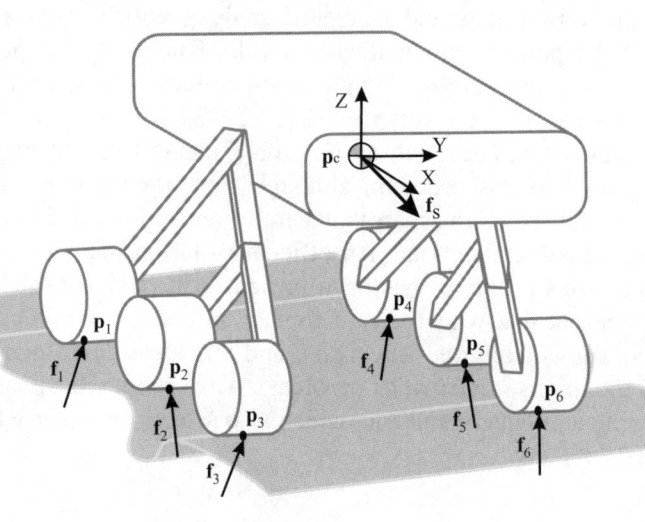

Fig. 2.3. Force analysis of a six-wheeled robot in rough terrain

A set of quasi-static force balance equations for the six-wheeled robot shown in Fig. 2.3 can be written as:

$$\begin{bmatrix} \mathbf{I} & \cdots & \mathbf{I} \\ \hline 0 & p_1^z & p_1^y & & 0 & p_6^z & p_6^y \\ -p_1^z & 0 & -p_1^x & \cdots & -p_6^z & 0 & -p_6^x \\ -p_1^y & p_1^x & 0 & & -p_6^y & p_6^x & 0 \end{bmatrix} \begin{bmatrix} \mathbf{f}_1 \\ \vdots \\ \mathbf{f}_6 \end{bmatrix} = \mathbf{f}_s \qquad (2.7)$$

where $\mathbf{I}$ represents a 3 x 3 identity matrix. This set of equations can be written in matrix form as:

$$\mathbf{G}\mathbf{x} = \mathbf{f}_s \qquad (2.8)$$

where $\mathbf{x} = [\mathbf{f}_1\ \mathbf{f}_2\ \mathbf{f}_3\ \mathbf{f}_4\ \mathbf{f}_5\ \mathbf{f}_6]^T$.

The solution to Eq. (2.8) will be discussed in detail in Chap. 4, in the context of motion control of robotic systems on uneven terrain. Here, we emphasize that the solution to Eq. (2.8) relies on an understanding of wheel-terrain interaction. Specifically, note that the wheel-terrain contact force vectors $\mathbf{f}_i$ can be decomposed into a tractive force T_i tangent to the wheel-terrain contact plane, a normal force N_i normal to the wheel-terrain contact plane, and a lateral force L_i in the wheel-terrain contact plane parallel to the wheel axis of rotation (see Fig. 2.4). T_i and N_i are functions of the applied wheel torque, which are the controllable inputs to the robot system. (L_i is generally uncontrollable, except for vehicles with extensible axles.) Therefore, to pose the force analysis problem, the orientation of the wheel-terrain contact planes, here termed the wheel-terrain contact angles, must be known. The solution to this problem will be discussed later in this chapter.

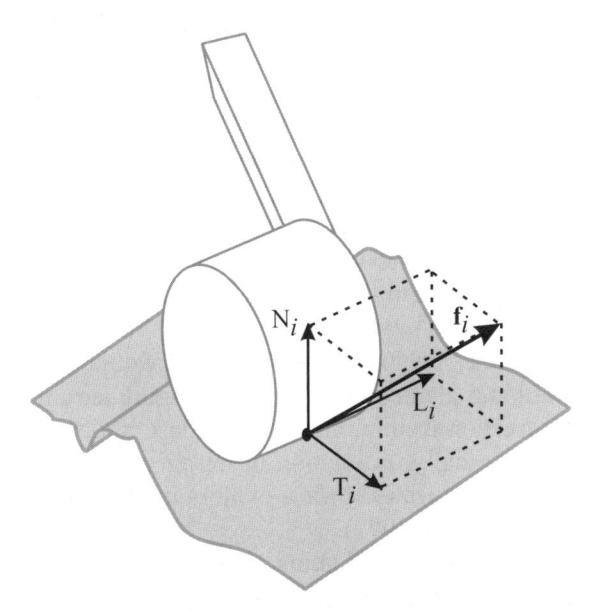

Fig. 2.4. Decomposition of wheel-terrain contact force vector

We also emphasize that the solution of the force analysis problem must obey physical constraints, including the constraint that the tractive force must not exceed the maximum shear force that the terrain can bear. The maximum terrain shear force is a function of the terrain cohesion and internal friction angle, and can computed from Coulomb's equation:

$$\tau_{max} = c + \sigma_{max} \tan\phi \qquad (2.9)$$

where c is the terrain cohesion and ϕ is the internal friction angle.

If the applied tractive pressure exceeds τ_{max}, terrain failure and excessive wheel slip may occur, which will greatly reduce the wheel traction. Thus, to predict τ_{max} and properly constrain the force analysis problem, c and ϕ must be known. A method for estimation of these parameters will be presented later in this chapter.

In summary, kinematic and force analyses are important elements of rough terrain motion planning and control algorithms. To pose these analyses, key terrain parameters and the wheel-terrain contact angles must be known. Methods for sensor-based estimation of these quantities are discussed below.

2.3 Terrain Characterization and Identification

The purpose of terrain characterization and identification is to identify key terrain parameters for use in wheel-terrain interaction models. This will enable accurate terrain traversability prediction.

Wheel-terrain interaction plays a critical role in rough terrain mobility (Bekker 1956, 1969; Wong 1976). Numerous researchers have developed models of wheel-terrain interaction that rely on physical terrain parameters that must be measured or estimated. It is important to estimate these parameters on-line, since this would allow a robot to predict its ability to safely traverse terrain (Iagnemma et al. 2003). It would also allow a robot to adapt its control and planning strategy to changing terrain conditions, to maximize wheel traction or minimize power consumption (Iagnemma and Dubowsky 2000b). Finally, estimation of terrain parameters is an important scientific goal of planetary surface exploration missions (Moore et al. 1977, 1999; Matijevic et al. 1997).

Here a method for on-line estimation of two key terrain parameters, cohesion, c, and internal friction angle, ϕ, is presented (Iagnemma et al. 2002). These parameters can be used to compute the maximum terrain shear strength, τ_{max}, from Coulomb's equation (Eq. (2.9)). Since soil failure occurs when the maximum shear strength is exceeded, knowledge of c and ϕ can be used to predict robot traversability on flat and sloped terrain.

To estimate terrain parameters, equations relating the parameters of interest to physically measurable quantities must be developed. A free-body diagram of a driven rigid wheel of radius r and width b traveling through deformable terrain is shown in Fig. 2.5. A vertical load W and a horizontal drawbar pull force DP are applied to the wheel by the vehicle suspension. A torque T is applied at the wheel rotation axis by an actuator. The wheel has angular velocity ω, and the wheel center has linear velocity, V. The angle from the vertical at which the wheel first makes contact with the terrain is denoted θ_1. The angle from the vertical at which the wheel loses contact with the terrain is denoted θ_2. Thus, the entire angular wheel-terrain contact region is defined by $\theta_1+\theta_2$.

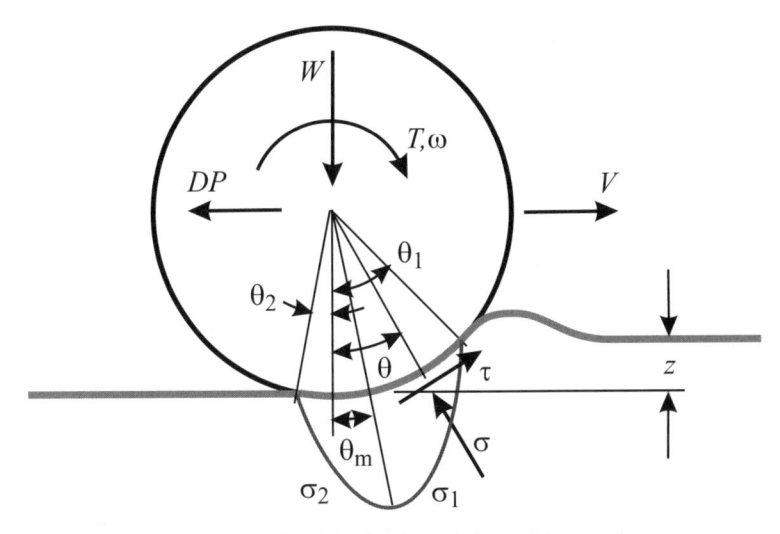

Fig. 2.5. Free-body diagram of rigid wheel on deformable terrain

A stress region is created at the wheel-terrain interface, as indicated by the regions σ_1 and σ_2. Here σ_1 is the interface section from initial terrain contact (i.e. θ_1) to the point of maximum stress (i.e. θ_m), and σ_2 is the region from the point of maximum stress to final terrain contact (i.e. θ_2). At a given point on the interface, the stress can be decomposed into a component acting normal to the wheel at the wheel-terrain contact point, σ, and a component acting tangential to the wheel at the wheel-terrain contact point, τ. The angle from the vertical at which the maximum stress occurs is denoted θ_m. Here we will assume $\theta_2 = 0$, since θ_2 is generally small in practice.

In the following analysis it is assumed that the following quantities are known or can be sensed or estimated: the vertical load W, torque T, sinkage z, wheel angular speed ω, and wheel linear speed V. Issues related to sensing of these quantities are discussed later in this section.

Force balance equations for the system in Fig. 2.5 can be written as:

$$W = rb\left(\int_{\theta_1}^{\theta_2} \sigma(\theta)\cos\theta \cdot d\theta + \int_{\theta_1}^{\theta_2} \tau(\theta)\sin\theta \cdot d\theta \right), \tag{2.10}$$

$$DP = rb\left(\int_{\theta_1}^{\theta_2}\tau(\theta)\cos\theta \cdot d\theta - \int_{\theta_1}^{\theta_2}\sigma(\theta)\sin\theta \cdot d\theta\right), \qquad (2.11)$$

$$T = r^2 b \int_{\theta_1}^{\theta_2}\tau(\theta)\cdot d\theta. \qquad (2.12)$$

The shear stress can be computed as:

$$\tau(\theta) = (c + \sigma(\theta)\tan\phi)\left(1 - e^{-\frac{r}{k}[\theta_1 - \theta - (1-i)(\sin\theta_1 - \sin\theta)]}\right) \qquad (2.13)$$

where k is the shear deformation modulus, r is the wheel radius, and i is the wheel slip, defined as $i = 1 - (V/r\omega)$ for $\omega > 0$ (Wong 1989).

The normal stress at the wheel-terrain interface is given by:

$$\sigma(z) = \left(\frac{k_c}{b} + k_\phi\right) z^n \qquad (2.14)$$

where k_c and k_ϕ are pressure-sinkage moduli, and n is the sinkage exponent (Bekker 1956). This equation can be expressed as a function of the wheel angular location θ by noting that sinkage is related to θ as:

$$z(\theta) = r(\cos\theta - \cos\theta_1). \qquad (2.15)$$

Substituting Eq. (2.15) into Eq. (2.16) yields expressions for the normal stress distribution along the wheel-terrain interface, as:

$$\sigma_1(\theta) = \left(\frac{k_c}{b} + k_\phi\right)(r(\cos\theta - \cos\theta_1))^n, \qquad (2.16)$$

$$\sigma_2(\theta) = \left(\frac{k_c}{b} + k_\phi\right)\left(r\left(\cos\left(\theta_1 - \theta\frac{(\theta_1 - \theta_m)}{\theta_m}\right) - \cos\theta_1\right)\right)^n. \qquad (2.17)$$

To develop a parameter estimation algorithm, analytical expressions for the force balance equations (Eqs. (2.10-2.12)) are required, since these equations relate physically measurable quantities (W, T, z, ω, V) to the parameters of interest (c, ϕ). However, Eqs. (2.10-2.12) are not amenable to closed-form integration, due to their complexity. This motivates the use of an approximate form of the stress equations (Eqs. (2.13, 2.16-2.17)).

2.3.1 Equation Simplification

Fig. 2.6 shows typical simulated plots of shear and normal stress distributions (as defined by Eqs. (2.13) and (2.16-2.17), respectively) around the rim of a driven rigid wheel on various terrains at moderate wheel slip. Parameters used in these plots are listed in Table 2.1, and represent a range of terrain types: dry sand, sandy loam, clayey soil, and snow (Wong 1976, 1989). It can be seen in Fig. 2.6 that the shear and normal stress distributions are approximately linear for a diverse range of terrains (Vincent 1961).

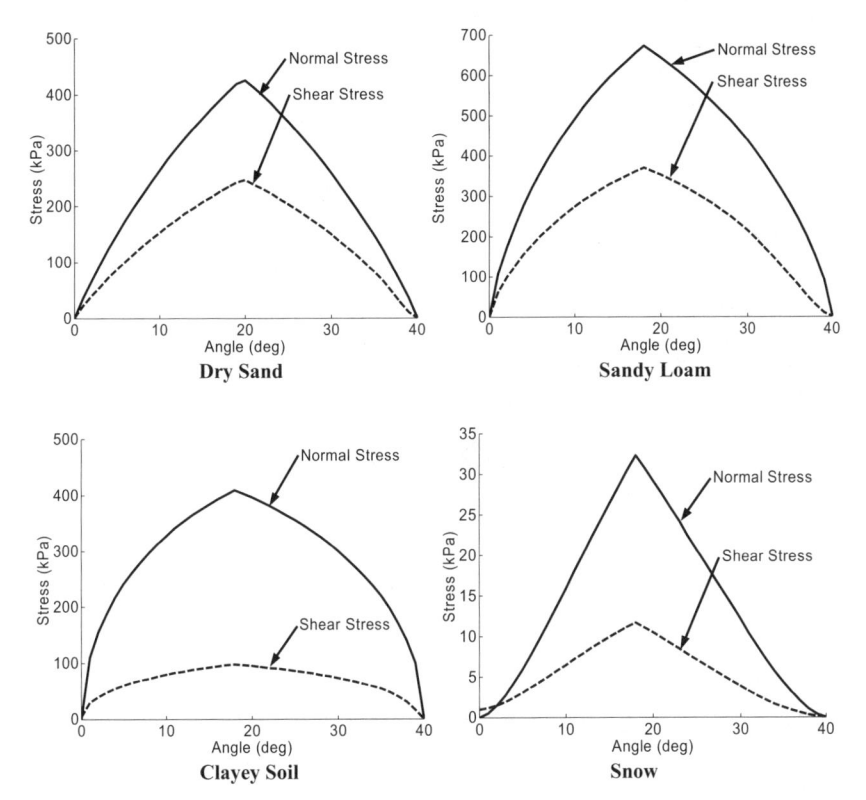

Fig. 2.6. Normal and shear stress distribution for various terrain types at moderate wheel slip

Based on this observation, linear approximations of the shear and normal stress equations can be written as:

$$\sigma_1^L(\theta) = \frac{\theta_1 - \theta}{\theta_1 - \theta_m} \sigma_m, \qquad (2.18)$$

$$\sigma_2^L(\theta) = \frac{\theta}{\theta_m} \sigma_m, \qquad (2.19)$$

$$\tau_1^L(\theta) = \frac{\theta_1 - \theta}{\theta_1 - \theta_m} \tau_m, \qquad (2.20)$$

$$\tau_2^L(\theta) = c + \frac{\theta}{\theta_m}(\tau_m - c), \qquad (2.21)$$

where σ_m and τ_m are the maximum values of the normal and shear stress, respectively. Eq. (2.21) contains an offset term for cohesive soils.

Table 2.1. Parameters for various terrain types

	Dry sand	Sandy loam	Clayey soil	Snow
n	1.1	0.7	0.5	1.6
c [kPa]	1.0	1.7	4.14	1.0
ϕ [deg]	30.0	29.0	13.0	19.7
k_c [kPa/m^{n-1}]	0.9	5.3	13.2	4.4
k_ϕ [kPa/m^n]	1523.4	1515.0	692.2	196.7
k [m]	0.025	0.025	0.01	0.04

Simplified forms of the force balance equations can be written by combining Eqs. (2.10-2.12) with Eqs. (2.18-2.21):

$$W = rb \begin{pmatrix} \int_0^{\theta_m} \sigma_2^L(\theta)\cos\theta \cdot d\theta + \int_{\theta_m}^{\theta_1} \sigma_1^L(\theta)\cos\theta \cdot d\theta \\ + \int_0^{\theta_m} \tau_2^L(\theta)\sin\theta \cdot d\theta + \int_{\theta_m}^{\theta_1} \tau_1^L(\theta)\sin\theta \cdot d\theta \end{pmatrix}, \qquad (2.22)$$

$$DP = rb \begin{pmatrix} \int_0^{\theta_m} \tau_2^L(\theta)\cos\theta \cdot d\theta + \int_{\theta_m}^{\theta_1} \tau_1^L(\theta)\cos\theta \cdot d\theta \\ - \int_0^{\theta_m} \sigma_2^L(\theta)\sin\theta \cdot d\theta - \int_{\theta_m}^{\theta_1} \sigma_1^L(\theta)\sin\theta \cdot d\theta \end{pmatrix}, \qquad (2.23)$$

$$T = r^2 b \left(\int_0^{\theta_m} \tau_2^L(\theta) \cdot d\theta + \int_{\theta_m}^{\theta_1} \tau_1^L(\theta) \cdot d\theta \right). \qquad (2.24)$$

2.3 Terrain Characterization and Identification

Evaluation of Eqs. (2.22) and (2.24) yields the following expressions for the normal load and torque:

$$W = \frac{rb}{\theta_m(\theta_1 - \theta_m)} \begin{pmatrix} \sigma_m(-\theta_m \cos\theta_1 + \theta_1 \cos\theta_m - \theta_1) \\ -\tau_m(\theta_m \sin\theta_1 - \theta_1 \sin\theta_m) \\ -c(\theta_1 \sin\theta_m - \theta_m \sin\theta_m - \theta_m\theta_1 + \theta_m^2) \end{pmatrix}, \quad (2.25)$$

$$T = \frac{r^2 b}{2}(\tau_m \theta_1 + c\theta_m). \quad (2.26)$$

Two assumptions are made in solving Eqs. (2.25) and (2.26) for c and ϕ. The first is that the location of the maximum shear and normal stress occurs at the same location, θ_m. Analysis and simulation have shown that this assumption is reasonable for a wide range of soil types. With this assumption, an additional relation can be written, based on Eq. (2.13):

$$\tau_m = (c + \sigma_m \tan\phi)\left(1 - e^{-\frac{r}{k}[\theta_1 - \theta_m - (1-i)(\sin\theta_1 - \sin\theta_m)]}\right). \quad (2.27)$$

The second assumption is that the angular location of maximum stress, θ_m, occurs midway between θ_1 and θ_2, i.e.:

$$\theta_m = \frac{\theta_1 + \theta_2}{2}. \quad (2.28)$$

This assumption is reasonable for a wide range of soils at moderate slip ratios (Kang 2003). This can be justified by noting that θ_m can be estimated from the relation $\theta_m = (c_1 + ic_2)\theta_1$, where c_1 and c_2 are terrain parameters. The range of c_1 and c_2 is generally $c_1 \approx 0.4$ and $0 \leq c_2 \leq 0.3$ (Wong and Reece 1967). Thus for a wide range of slip ratios, θ_m will be near 0.5.

The system of Eqs. (2.25-2.27) can be combined into a single equation relating cohesion and internal friction angle, as:

$$c = \frac{\kappa_1 \tan\phi + \kappa_2}{\kappa_3 \tan\phi + \kappa_4} \quad (2.29)$$

where:

$$\kappa_1 = A\left(\theta_1^2 Wr + 4T\sin\theta_1 - 8T\sin\frac{\theta_1}{2}\right)$$

$$\kappa_2 = 4T\left(\cos\theta_1 - 2\cos\frac{\theta_1}{2} + 1\right)$$

$$\kappa_3 = A\theta_1 r^2 b\left(\sin\theta_1 - 4\sin\frac{\theta_1}{2} + \theta_1\right)$$

$$\kappa_4 = \theta_1 r^2 b\left(\cos\theta_1 - 2\cos\frac{\theta_1}{2} + 2A\cos\theta_1 - 4A\cos\frac{\theta_1}{2} + 2A + 1\right)$$

and $A = 1 - e^{-\frac{r}{k}\left[\frac{\theta_1}{2} + (1-i)\left(-\sin\theta_1 + \sin\left(\frac{\theta_1}{2}\right)\right)\right]}$.

Eq. (2.29) can be rearranged to the following form:

$$\frac{K_2}{K_4} = \frac{K_3}{K_4} c \tan\phi + c - \frac{K_1}{K_4} \tan\phi. \tag{2.30}$$

The relative contribution of each term in the right-hand side of Eq. (2.30) was studied numerically over the range of parameters in Table 2.2. This space encompasses a broad variety of terrain types. The simulated wheel radius r was 0.1 m, and the wheel width b was 0.1 m.

It was found that the maximum relative contribution of the $(K_3/K_4)c\tan\phi$ term was 2.47%. The other two terms had a significantly higher contribution (Kang 2003). Thus the $(K_3/K_4)c\tan\phi$ term is negligible, and Eq. (2.30) can be reduced to:

$$\frac{K_2}{K_4} = c - \frac{K_1}{K_4} \tan\phi. \tag{2.31}$$

Eq. (2.31) is a single equation in two unknowns. At least two unique instances of Eq. (2.31) are required to compute c and ϕ. During the parameter estimation process, it is expected that sensor data would be sampled at a frequency of several hertz. For each unique sensor sampling occurrence j, a unique instance of Eq. (2.31) can be written:

$$\begin{aligned}\frac{K_2^1}{K_4^1} &= c - \frac{K_1^1}{K_4^1}\tan\phi \\ &\vdots \\ \frac{K_2^j}{K_4^j} &= c - \frac{K_1^j}{K_4^j}\tan\phi\end{aligned} \tag{2.32}$$

or in matrix form:

$$\mathbf{K}_1 = \mathbf{K}_2 \begin{bmatrix} c \\ \tan\phi \end{bmatrix} \tag{2.33}$$

with $\mathbf{K}_1 = \begin{bmatrix} K_2^1/K_4^1 \ldots K_2^j/K_4^j \end{bmatrix}^T$, $\mathbf{K}_2 = \begin{bmatrix} 1 & \cdots & 1 \\ -K_1^1/K_4^1 & \cdots & -K_1^j/K_4^j \end{bmatrix}^T$.

Table 2.2. Parameter space for algorithm analysis

Minimum value	Parameter	Maximum value
0.5	n	1.2
20.0	ϕ [deg]	40.0
0.0	c [kPa]	10.0
10.0	k_c [kPa/m^{n-1}]	100.0
1000.0	k_ϕ [kPa/m^n]	5000.0
0.01	k [m]	0.03

In practice, more than two equations are used to form an estimate of c and ϕ to decrease sensitivity to sensor noise. In this case $\mathbf{K}_2$ is nonsquare and Eq. (2.33) can be solved in a least-squares sense, using the pseudoinverse of $\mathbf{K}_2$:

$$\begin{bmatrix} c \\ \tan\phi \end{bmatrix} = \left(\mathbf{K}_2^T \mathbf{K}_2\right)^{-1} \mathbf{K}_2^T \mathbf{K}_1. \tag{2.34}$$

Note that singularity of $(\mathbf{K}_2^T \mathbf{K}_2)^{-1}$ only occurs in the degenerate case where non-unique sensor data is sampled (such as on perfectly flat terrain with noise-free sensors). This case will be discussed later.

All quantities in Eq. (2.34) can be sensed except the shear deformation modulus k (in the matrices $\mathbf{K}_1$ and $\mathbf{K}_2$). In practice, the estimation algorithm exhibits low sensitivity to k, particularly for large wheel radii and high slip ratios. Therefore k is usually chosen as a representative value for deformable terrain. Techniques for estimating k are described in (Kang 2003).

2.3.2 Sensing and Implementation Issues

In the preceding analysis it was assumed that the vertical load W, torque T, sinkage z, wheel angular speed ω, and wheel linear speed V could be measured or estimated. Here, methods for measuring or estimating these inputs are discussed, along with other implementation issues for planetary rovers.

The vertical load W can be computed from a quasi-static force analysis of the robot, with knowledge of the robot configuration and mass distribution. Quasi-static analysis is valid since dynamic effects are negligible at the low speeds of these vehicles (Weisbin et al. 1999). The torque T can be estimated from the current input to the motor and an empirically-determined mapping from current to torque. In applications where large thermal variation is expected (such as Martian surface exploration), motor temperature can be included in this mapping (Matijevic et al. 1997). Note that torque and vertical load could be directly measured if the wheel were instrumented with a multi-axis force sensor. However, this adds cost and complexity to the robot.

The sinkage z can be computed with vision-based techniques or by kinematic analysis of the robot suspension (Wilcox 1994; Iagnemma et al. 2003). The wheel angular speed ω can be measured with a tachometer. The wheel linear speed V can be computed using inertial measurement unit (IMU) measurements. However, at low speeds IMU velocity measurements can be highly degraded by noise. In this case, visual odometry can yield more accurate results (Olson et al. 2001).

The sensors described above (i.e. robot configuration sensors, motor current sensor, wheel tachometer, IMU, and vision system) would be part of most autonomous robot systems. Thus, all required inputs can be measured or estimated with on-board robot sensors.

An important implementation issue is minimizing sensor noise. Most robot sensors can be modeled as a "true" signal corrupted by white noise. In this case, increasing the number of data points in Eq. (2.34) acts as an averaging filter and improves estimation accuracy. Other filtering techniques (such as the Kalman filter and its extensions) could also be applied to this problem.

Note that all data points used in a parameter estimate are assumed to be sampled from homogeneous terrain. For example, consider a robot moving at 5 cm/sec with a sensor sampling rate of 5 Hz. If 10 data points are used to compute a parameter estimate, it must be assumed that the terrain is homogeneous within a 10 cm distance. If data is sampled from mixed or inhomogeneous terrain, resulting parameter estimates will be effective estimates of the combined terrain types. In general, assumptions regarding terrain homogeneity can be formed from *a priori* knowledge of local terrain characteristics. Terrain classification methods could also be used to detect changes in terrain type (Belluta et al. 2000).

A final implementation issue arises for a robot traveling at constant velocity on flat terrain, where the matrix $(\mathbf{K}_2^T \mathbf{K}_2)^{-1}$ in Eq. (2.34) may be poorly conditioned. This occurs because the robot is collecting an identical set of sensor readings at each sampling instance. The ridge regression technique can be used to solve Eq. (2.34) in cases where $(\mathbf{K}_2^T \mathbf{K}_2)^{-1}$ is poorly conditioned as:

$$\begin{bmatrix} c \\ \tan \phi \end{bmatrix} = \left(\mathbf{K}_2^T \mathbf{K}_2 + \delta \mathbf{I} \right)^{-1} \mathbf{K}_2^T \mathbf{K}_1 \qquad (2.35)$$

where δ is a small positive constant that can be optimized by techniques such as cross-validation (Golub and van Loan 1996). In practice (and in the results presented below), natural terrain variation will usually lead to acceptable equation conditioning. Deliberately inducing variable wheel slip also improves parameter estimates on flat terrain.

2.4 Results: Terrain Identification

2.4.1 Simulation Results

Dynamic simulations were conducted of a single driven wheel traveling through deformable terrain. The purpose was to examine algorithm accuracy under various terrain conditions. In all simulations the wheel traveled at approximately 10 cm/sec. The wheel had radius 0.1 m, width 0.1 m, mass 10 kg, and inertia 0.05 kg·m^2. A proportional-derivative control algorithm commanded the wheel. The simulated sampling rate was 20 Hz. Further details of the simulation are given in (Kang 2003).

Simulations were first performed with noise-free inputs W, T, z, and i, to study fundamental algorithm accuracy. Random variation of 15% was introduced to W and i to simulate variation caused by natural terrain unevenness, where wheel load and velocity change as the robot configuration changes. During the estimation procedure, the shear deformation modulus k was assumed to be 150% of its actual value to study algorithm sensitivity to this parameter.

Estimates of cohesion and internal friction angle were computed for 15,625 evenly-spaced parameter sets in the parameter space described in Table 2. This parameter space represents a broad variety of terrain types. Five sampled data points were used to compute each parameter estimate.

The RMS error between the estimated and actual c over all simulations was 0.21 kPa (c ranges from 0-10 kPa). The RMS error between the estimated and actual ϕ over all simulations was 1.62° (ϕ ranges from 20°-40°). This shows that the approximations introduced in Sect. 2.3.1 do not introduce significant error into the estimation algorithm. The computational load for the algorithm was approximately 1 msec per estimation cycle for unoptimized Matlab code on a 933 MHz desktop PC.

Simulations were then run to study the effect of sensor noise on estimation accuracy. The inputs W, T, z, and i were corrupted with white noise with standard deviation equal to 10% of their maximum value. Again, estimates of cohesion and internal friction angle were computed for the 15,625 parameter sets described above. The shear deformation modulus k was assumed to be 150% of its actual value, and 10 sampled data points were used to compute each parameter estimate.

The RMS error between the estimated and actual c over all simulations was 1.57 kPa. The RMS error between the estimated and actual ϕ over all simulations was 4.12°. As expected, increasing the sensor noise levels leads to increased parameter estimation error. Simulations were then run with the number of sampled data points increased to 30. The RMS error between the estimated and actual c was reduced to 0.63 kPa, and for ϕ to 2.11°. The computational load for the algorithm was approximately 3 msec per estimation cycle. This suggests that it is possible to compute accurate parameter estimates despite sensor noise, with low computational cost. It also shows that increasing the number of sampled data points used to form a parameter estimate decreases the error caused by sensor noise.

Figs. 2.7 and 2.8 show representative parameter estimation simulation results for dry sand (see Table 2.1). It can be seen that the estimated parameters c and ϕ rapidly approach their true values in the noise-free case. Sensor noise degrades the accuracy of the estimated parameters compared to the noise-free case. The effect of sensor noise is diminished by increasing the number of sampled data points from 15 to 30. In general it is possible to obtain accurate parameter estimates using relatively few data samples despite sensor noise.

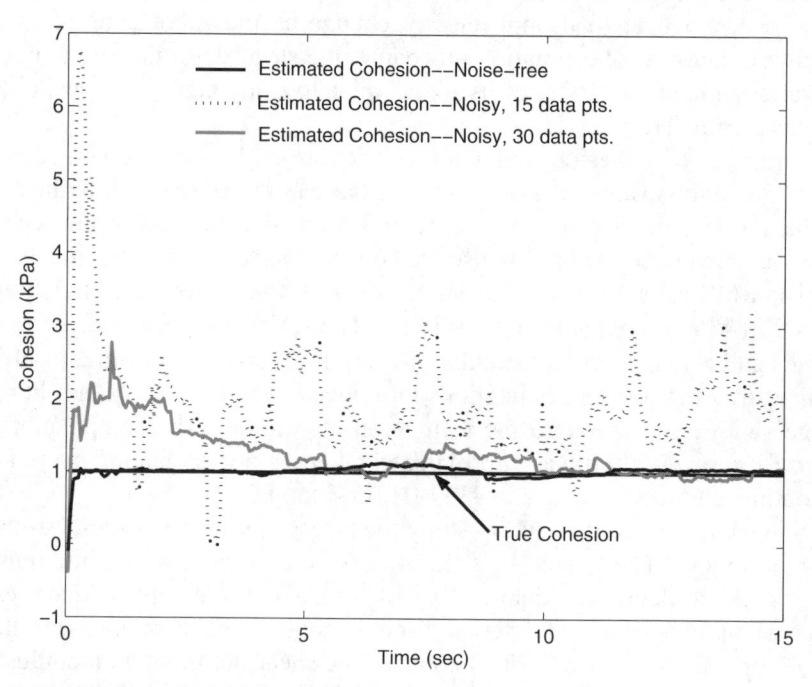

Fig. 2.7. Simulated estimation of cohesion of dry sand, for noise-free case, noisy case using 15 data points, and noisy case using 30 data points

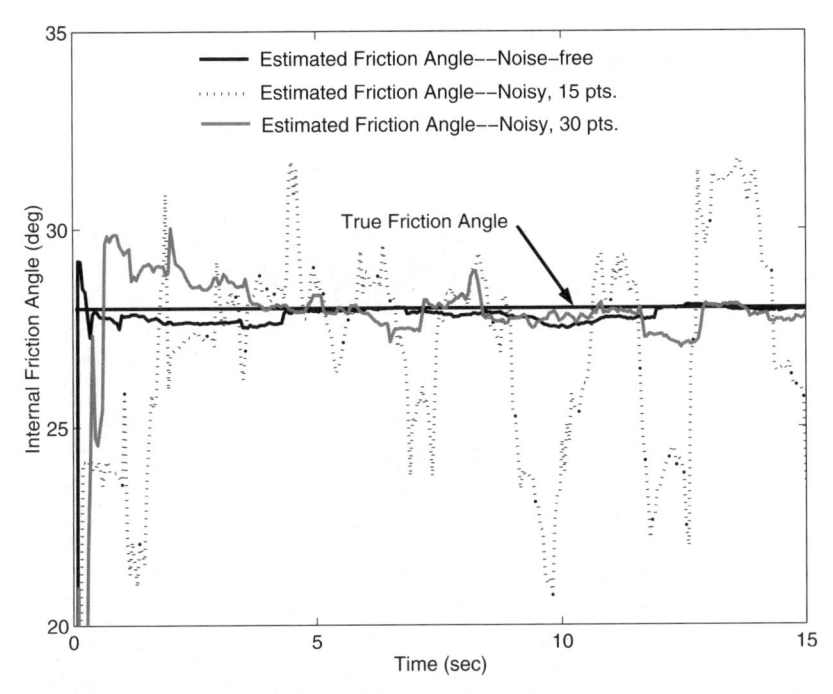

Fig. 2.8. Simulated estimation of internal friction angle of dry sand, for noise-free case, noisy case using 15 data points, and noisy case using 30 data points

2.4.2 Experimental Results

Experiments were performed on a laboratory terrain characterization testbed, shown in Fig. 2.9. The testbed consists of a driven rigid wheel mounted on an undriven vertical axis. The wheel assembly is mounted to a driven horizontal carriage. By driving the wheel and carriage at different rates, variable slip ratios can be imposed. The vertical wheel load can be changed by adding weight to the vertical axis.

The testbed is instrumented with encoders to measure angular velocities of both the wheel and the carriage pulley. The carriage linear velocity is computed from the carriage pulley angular velocity. The vertical wheel sinkage is measured with a linear potentiometer. The wheel torque T is measured by a Cooper Instruments torque sensor. The six-component wrench between the wheel and carriage is measured with a JR3 six-axis force/torque sensor. The force sensor allows measurement of the normal load W and drawbar pull DP. The testbed is controlled by a 133 MHz PC. The soil bin is 90 cm long, 30 cm wide and 15 cm deep.

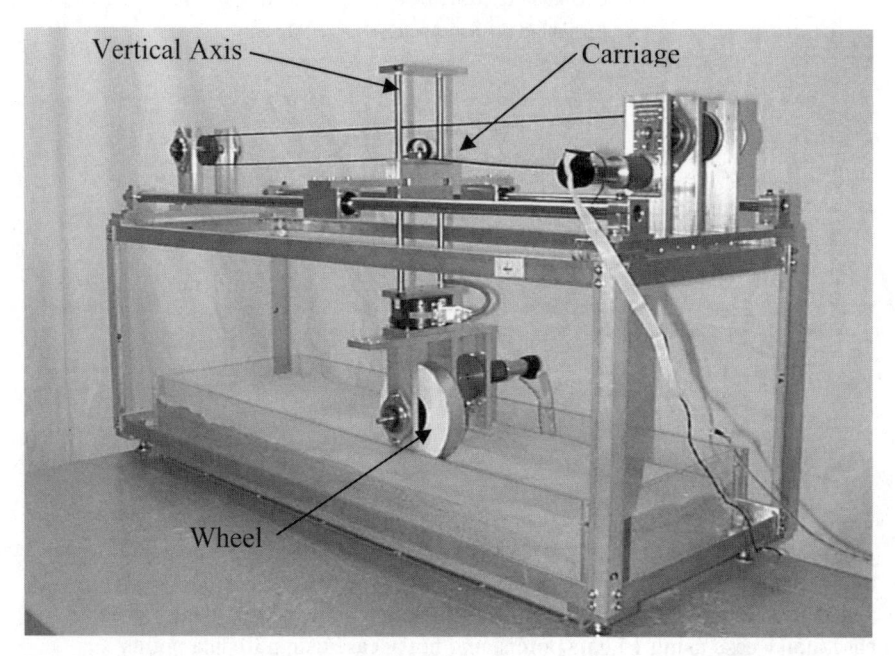

Fig. 2.9. Terrain characterization testbed

The wheel diameter and width are 14.6 and 6.0 cm, respectively. The wheel maximum angular velocity is 1.1 rad/sec. This results in a maximum linear velocity of 8.0 cm/sec. The wheel size and speed were chosen to be in the range of current and projected planetary rovers.

Three distinct soil types were chosen for experimentation: washed sand, dried bentonite clay, and compacted topsoil. Classical shear failure experiments were performed to determine baseline values of c and ϕ for all soils. In these experiments, a vertical load is applied to homogeneous soil through a device called a bevameter, which is translated horizontally until shear failure occurs (see Fig. 2.10) (Bekker 1956). By varying the vertical load, a linear relationship between normal stress and shear stress can be observed, and c and ϕ can be estimated. Note that this is an off-line method of terrain identification and is not suitable for on-line characterization.

Numerous experiments were run to account for nonuniformity in soil mixing and moisture content. Table 3 summarizes the results of these experiments. These results agree with published values for similar soils

(Wong 1976, 1989; Yong et al. 1984). A result of shear failure experiments for dried bentonite clay can be seen in Fig. 2.11. This figure shows a linear relationship between normal stress and shear stress, as predicted by Eq. (2.9).

Fig. 2.10. Bevameter (BEkker VAlue METER) for soil parameter identification

Parameter identification experiments were then performed on each soil type. The shear deformation modulus k was assumed to be 0.05. Thirty sampled data points were used to compute each parameter estimate.

Fig. 2.12 shows the results of the estimation algorithm for dried bentonite clay. The estimated cohesion and internal friction angle rapidly converge to values of approximately 0.70 kPa and 32.1°, respectively. These values lie near or within the variance observed in the shear failure experiments.

Table 2.4 summarizes the results of these experiments. Comparing these values to the values of Table 2.3, it can be seen that the estimated values are similar to those measured through shear failure experiments. This suggests that the proposed approach can identify c and ϕ of various soils despite noisy sensors. Error and variation in estimated parameters is likely due to nonuniformity in soil mixing and moisture content. Note that this variation is not unique to the proposed method, but is present in any soil property measurement technique. Estimation error is also due to sensor noise and error in the assumed value of k. Additionally, at high slip ratios the testbed wheel exhibited control chatter, which degraded sensor readings.

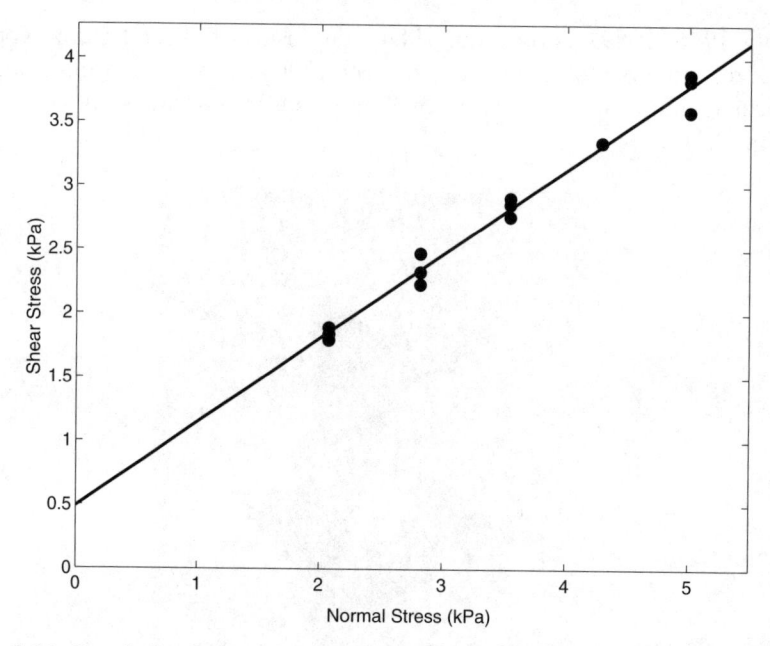

Fig. 2.11. Result from shear failure experiments for dried bentonite clay. Data points shown with best-fit line

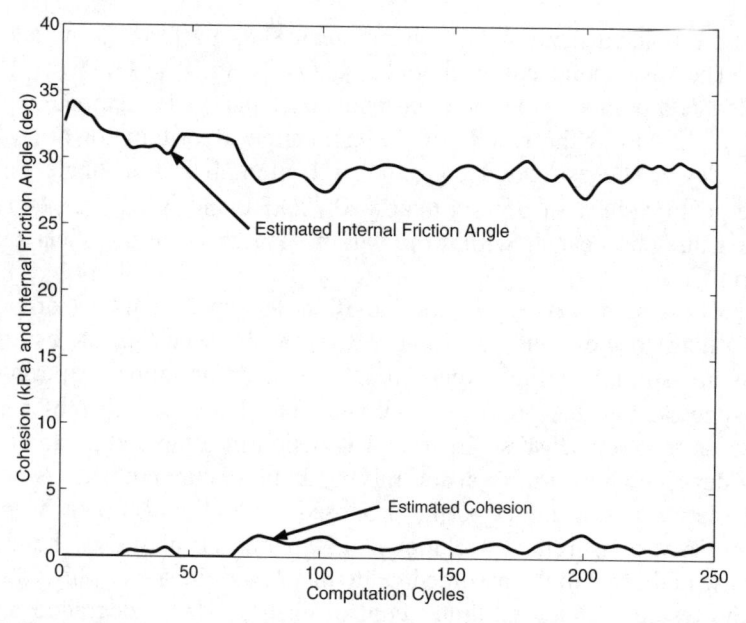

Fig. 2.12. Experimental cohesion and internal friction angle estimation for dried bentonite

Table 2.3. Results from shear failure experiments

Terrain Type	c [kPa]	ϕ [deg]
Washed Sand	0.65 ± 0.24	32.1 ± 2.82
Dried Bentonite Clay	0.48 ± 0.19	33.7 ± 1.99
Compacted Topsoil	0.74 ± 0.24	44.3 ± 2.01

Table 2.4. Results from terrain parameter estimation experiments

	c [kPa]	ϕ [deg]
Washed Sand	0.77 ± 0.48	29.6 ± 1.47
Dried Bentonite Clay	0.70 ± 0.35	32.1 ± 2.60
Compacted Topsoil	1.04 ± 0.43	43.7 ± 3.11

These results show that the estimation algorithm produces reasonably accurate, on-line parameter estimates on an experimental system with noisy sensors, for a variety of terrain types. The level of accuracy shown by the algorithm would allow it to distinguish between distinct soil types such as crusty material, drift material, clay, etc. The computation time for each estimation cycle was approximately 1 msec on a 933 MHz desktop PC. Thus the approach is suitable for systems with limited on-board computation, such as planetary rovers.

2.5 Wheel-Terrain Contact Angle Estimation

The purpose of wheel-terrain contact angle estimation is to estimate the angle from the horizontal of the point of contact of the robot wheel and the terrain. This enables terrain traversability prediction, as described earlier, and is useful for motion planning algorithms discussed later in this monograph. Here, a method for estimating wheel-terrain contact angles from on-board robot sensors is presented.

Consider a planar two-wheeled system on uneven terrain, as shown in Fig. 2.13. A planar analysis is appropriate since most wheeled vehicles can neither move instantaneously nor apply forces in the transverse direction. Thus, transverse contact angles are not considered. In this analysis the terrain is treated as rigid, and the wheels are assumed to make point contact with the terrain.

In Fig. 2.13 the rear and front wheels make contact with the terrain at angles γ_1 and γ_2 from the horizontal, respectively. The vehicle pitch, α, is also defined with respect to the horizontal. The wheel centers have speeds v_1 and v_2. These speeds are in a direction parallel to the local wheel-terrain tangent due to the rigid terrain assumption. The distance between the wheel centers is defined as l.

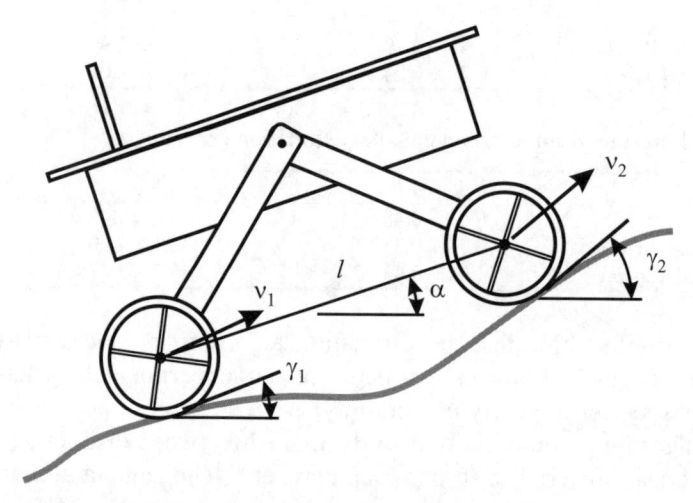

Fig. 2.13. Planar two-wheeled system

For this system, the following kinematic equations can be written:

$$v_1 \cos(\gamma_1 - \alpha) = v_2 \cos(\gamma_2 - \alpha), \qquad (2.36)$$

$$v_2 \sin(\gamma_2 - \alpha) - v_1 \sin(\gamma_1 - \alpha) = l\dot{\alpha}. \qquad (2.37)$$

Eq. (2.36) represents the kinematic constraint that the wheel center length l does not change. Note that this constraint remains valid in cases where changes in the vehicle suspension configuration cause changes in l, as long as l varies slowly. Eq. (2.37) is a rigid-body kinematic relation between the velocities of the wheel centers and the vehicle pitch rate $\dot{\alpha}$.

Combining Eqs. (2.36) and (2.37) yields:

$$\sin(\gamma_2 - \alpha - (\gamma_1 - \alpha)) = \frac{l\dot{\alpha}}{v_1} \cos(\gamma_2 - \alpha). \qquad (2.38)$$

With the definitions:

$$\theta \equiv \gamma_2 - \alpha, \quad \beta \equiv \alpha - \gamma_1, \quad a \equiv l\dot{\alpha}/v_1, \quad b \equiv v_2/v_1.$$

Eqs. (2.36) and (2.38) become:

$$(b\sin\theta + \sin\beta)\cos\theta = a\cos\theta, \qquad (2.39)$$

$$\cos\beta = b\cos\theta. \qquad (2.40)$$

2.5 Wheel-Terrain Contact Angle Estimation

Solving Eqs. (2.39) and (2.40) for the wheel-terrain contact angles γ_1 and γ_2 yields:

$$\gamma_1 = \alpha - \cos^{-1}(h), \qquad (2.41)$$

$$\gamma_2 = \cos^{-1}(h/b) + \alpha \qquad (2.42)$$

where $h \equiv \dfrac{1}{2a}\sqrt{2a^2 + 2b^2 + 2a^2b^2 - a^4 - b^4 - 1}$.

There are two special cases that must be considered in this analysis. The first special case occurs when the robot is stationary. In this case Eqs. (2.39) and (2.40) do not yield a solution, since if $\dot{\alpha} = v_1 = v_2 = 0$, both a and b are undefined. Physically, the lack of a solution results from the fact that a stationary robot can have an infinite set of possible contact angles at each wheel.

The second special case occurs when $\cos(\theta)$ equals zero. In this case $\gamma_2 = \pm\pi/2 + \alpha$ from the definition of θ, and Eq. (2.42) yields the solution $\gamma_1 = \pm\pi/2 + \alpha$. Physically this corresponds to two possible cases: the robot undergoing pure translation or pure rotation, as shown in Fig. 2.14.

Pure Translation **Pure Rotation**

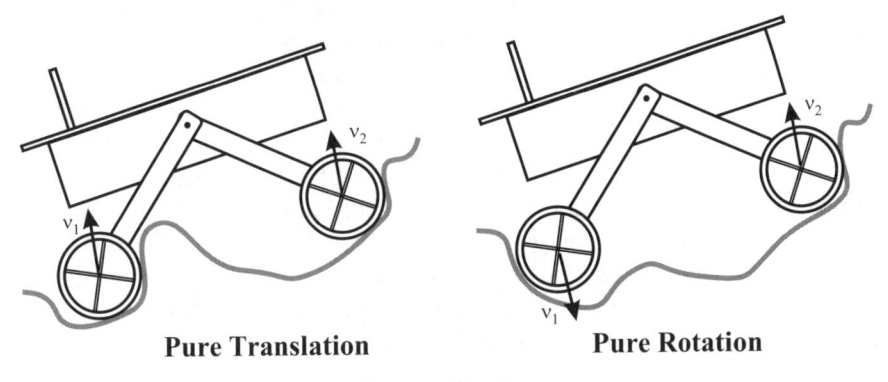

Fig. 2.14. Physical interpretations of $\cos(\theta) = 0$

While these cases are unlikely to occur in practice, they are easily detectable. For the case of pure rotation, $v_1 = -v_2$. The solutions for γ_1 and γ_2 can be written by inspection as:

$$\gamma_1 = \alpha + \frac{\pi}{2}\mathrm{sgn}(\dot{\alpha}), \qquad (2.43)$$

$$\gamma_2 = \alpha - \frac{\pi}{2}\mathrm{sgn}(\dot{\alpha}). \qquad (2.44)$$

For the case of pure translation, $\dot{\alpha} = 0$, and $v_1 = v_2$. Thus h is undefined and the system of Eqs. (2.39) and (2.40) has no solution. However, for low-speed robots considered in this work, the terrain profile varies slowly with respect to the data sampling rate. It is reasonable to assume that wheel-terrain contact angles computed at a given timestep will be similar to wheel-terrain contact angles computed at the previous timestep. Thus, previously estimated contact angles can be used when a solution to the estimation equations does not exist.

The pitch and pitch rate can be measured with rate gyroscopes or simple inclinometers. The wheel center speeds can be estimated from the wheel angular rate as measured by a tachometer, provided the wheels do not have substantial slip. Thus, wheel-terrain contact angles can be estimated with common, low-cost on-board sensors. The estimation process is computationally simple, and thus suitable for on-board implementation.

2.5.1 Extended Kalman Filter Implementation

The above analysis suggests that wheel-terrain contact angles can be computed from simple, measurable quantities. However, sensor noise and wheel slip will degrade these measurements. Here, an extended Kalman filter (EKF) is developed to compensate for these effects. This filter is an effective framework for fusing data from multiple noisy sensor measurements to estimate the state of a nonlinear system (Brown and Hwang 1997; Welch and Bishop 1999). In this case the sensor signals are wheel tachometers, gyroscopes, and inclinometers, and are assumed to be corrupted by unbiased Gaussian white noise with known covariance. Again, due to the assumption of quasi-static vehicle motion, inertial effects do not corrupt the sensor measurements. Also, the sensor bandwidth is assumed to be significantly faster than the vehicle dynamics, and thus sensor dynamics do not corrupt the sensor measurements.

Here the state vector $\mathbf{x}$ is estimated, composed of the wheel-terrain contact angles, i.e. $\mathbf{x} = [\gamma_1 \; \gamma_2]^T$. The discrete-time equation governing the evolution of $\mathbf{x}$ is:

$$\mathbf{x}_{k+1} = \begin{bmatrix} 1 & 0 \\ 0 & 1 \end{bmatrix} \mathbf{x}_k + w_k \qquad (2.45)$$

where w_k is a 2 x 1 vector representing the process noise. Eq. (2.45) implies that ground contact angle evolution is a Wiener process. This is physically reasonable, since terrain variation is inherently unpredictable. The process noise covariance can be assigned as the expected terrain variation:

2.5 Wheel-Terrain Contact Angle Estimation

$$\mathbf{R}_k = E\left((\mathbf{x}_{k+1} - \mathbf{x}_k)^2\right). \qquad (2.46)$$

This information could be estimated from knowledge of local terrain roughness, or computed from forward-looking range data.

The EKF measurement equation can be written as:

$$\mathbf{y}_k = \begin{bmatrix} 1 & 0 \\ 0 & 1 \end{bmatrix} \mathbf{x}_k + n_k \qquad (2.47)$$

where $\mathbf{y}_k$ is a synthetic "measurement" of the ground contact angles, computed analytically from Eqs. (2.41-2.42) and raw sensor data, i.e. $\mathbf{y}_k = f(\mathbf{z}_k)$, where $\mathbf{z} = [\alpha \; \dot{\alpha} \; v_1 \; v_2]^T$. It is assumed that the vehicle pitch α and pitch rate $\dot{\alpha}$ are directly sensed, and speeds v_1 and v_2 can be approximated from knowledge of the wheel angular velocities and radii.

The following is a description of the EKF implementation procedure:

1. Initialization of the state estimate $\hat{\mathbf{x}}_0$ and the estimated error covariance matrix $\mathbf{P}_0$. Here, $\hat{\mathbf{x}}_0 = \mathbf{y}_0$, and $\mathbf{P}_0 = \mathbf{R}_x$, where $\mathbf{R}_x = E\{w_0 w_0^T\}$.

2. Propagation of the current state estimate and covariance matrix. The state estimate is generally computed from a state transition matrix, which here is the identity matrix. Thus:

$$\hat{\mathbf{x}}_k^- = \hat{\mathbf{x}}_{k-1}^-. \qquad (2.48)$$

The *a priori* covariance matrix is computed as:

$$\mathbf{P}_k^- = \mathbf{P}_{k-1} + \mathbf{R}_x. \qquad (2.49)$$

3. Computation of the Kalman gain, and updating the *a priori* state estimate and a priori covariance matrix. The Kalman gain matrix $\mathbf{K}$ is given by:

$$\mathbf{K}_k = \mathbf{P}_k^- \left(\mathbf{P}_k^- + \mathbf{R}_{y_k}\right)^{-1}. \qquad (2.50)$$

The sensor noise covariance matrix $\mathbf{R}_{y_k}$ can be computed as:

$$\mathbf{R}_{y_k} = E\{n_k n_k^T\} = \left(\frac{\partial \mathbf{y}_k}{\partial \mathbf{z}}\right)^T \mathbf{R}_z \left(\frac{\partial \mathbf{y}_k}{\partial \mathbf{z}}\right). \qquad (2.51)$$

where $\mathbf{R}_z$ is a 4 x 4 diagonal matrix of known noise covariances associated with $\mathbf{z}$: $\mathbf{R}_z = diag(\sigma_\alpha^2, \sigma_{\dot{\alpha}}^2, \sigma_{v_1}^2, \sigma_{v_2}^2)$. Note that an estimate of $\mathbf{R}_{y_k}$ can also be formed by computing the unscented transform of Eqs. (2.41-2.42) (Julier and Uhlmann 1997).

The state estimate is updated as:

$$\hat{\mathbf{x}}_k = \hat{\mathbf{x}}_k^- + \mathbf{K}_k\left(\mathbf{y}_k - \hat{\mathbf{x}}_k^-\right) \tag{2.52}$$

and the covariance matrix is updated as:

$$\mathbf{P}_k = (\mathbf{I} - \mathbf{K}_k)\mathbf{P}_k^-. \tag{2.53}$$

The special cases discussed in Sect. 2.5 can lead to a lack of observability in the filter. However, as described above, these situations are detectable. Thus, new measurement updates for the filter are not taken when these special cases are detected.

A pictorial diagram of the EKF estimation process is shown in Fig. 2.15 (adapted from Welch and Bishop 1999).

Initial estimates for $\hat{\mathbf{x}}_0$ and $\mathbf{P}_0$ Measurement $\mathbf{y}_k = f(\mathbf{z}_k)$

Project state forward:
$$\hat{\mathbf{x}}_k^- = \hat{\mathbf{x}}_{k-1}^-$$
Project covariance forward:
$$\mathbf{P}_k^- = \mathbf{P}_{k-1} + \mathbf{R}_x$$

Compute Kalman gain:
$$\mathbf{K}_k = \mathbf{P}_k^-\left(\mathbf{P}_k^- + \mathbf{R}_{y_k}\right)^{-1}$$
Update state estimate with measurement:
$$\hat{\mathbf{x}}_k = \hat{\mathbf{x}}_k^- + \mathbf{K}_k\left(\mathbf{y}_k - \hat{\mathbf{x}}_k^-\right)$$
Update error covariance:
$$\mathbf{P}_k = (\mathbf{I} - \mathbf{K}_k)\mathbf{P}_k^-$$

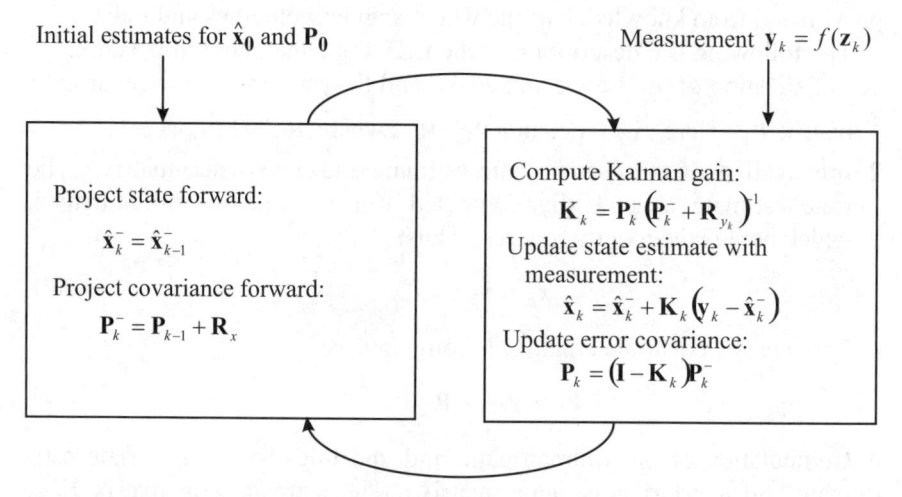

Fig. 2.15. Diagram of EKF estimation process (from Welch and Bishop 1999)

2.6 Results: Wheel-Terrain Contact Angle Identification

2.6.1 Simulation Results

The wheel-terrain contact angle estimation algorithm was implemented in simulation on a planar vehicle similar to the one in Fig. 2.13, with $l = 1$ m. The pitch α was corrupted with white noise of standard deviation 3°. The rear and front wheel velocities, v_1 and v_2, were corrupted with white noise of standard deviation 0.5 cm/sec. This models error due to effects of wheel slip and tachometer noise.

Fig. 2.16 shows the results of a representative simulation trial on undulating terrain. Here, the actual and estimated wheel-terrain contact angles are compared. It can be seen that after an initial transient, the EKF estimates of the terrain contact angles are quite accurate, with RMS errors of 0.80° and 0.81° for the front and rear angles, respectively. Error increases at flat terrain regions (i.e. where the values of front and rear contact angles are identical) since the angle estimation equations become poorly conditioned due to reasons discussed previously. However, the error covariance matrix remained small during the simulation. In general, the EKF does an excellent job in simulation of estimating wheel-terrain contact angles in the presence of noise.

2.6.2 Experimental Results

The wheel-terrain contact angle estimation algorithm was implemented on the six-wheeled experimental rover system shown in Fig. 2.17 (Burn 1997). This rover possesses a rocker-bogie suspension similar to the Sojourner rover (Bickler 1992). The six wheels are driven by geared DC motors with a peak torque of 100 oz-in and maximum angular velocity of 12 rpm. The resulting maximum velocity of the rover is approximately 8 cm/sec. The rover is steered with skid-steering. A mechanical differential in the rover frame allows the body to "split the difference" of the two rocker angles. The rover weighs 6.1 kg. The rover has on-board power and a PC/104 computer for autonomous control. Additional PC/104 modules support digital and analog IO, and sensor reading. A wireless modem is used for external communication.

The rover sensor suite is composed of tachometers to measure the wheel angular velocities, and a three-axis Crossbow CXL04M3 accelerometer mounted to the rover body to determine roll and pitch relative to an inertial frame. A JR3-67M25A six-axis force/torque sensor is mounted at the front of the rover, to measure forces exerted on the rover body by a three d.o.f. manipulator.

For this type of rocker-bogie suspension, the contact angles for the front and middle wheels can be estimated with Equations (2.41) and (2.42). The rear wheel does not maintain a fixed distance from either the front or middle wheel due to the rover's articulated suspension. However, it does maintain a fixed distance to the bogey free-pivot joint and its contact angle can be estimated from Equations (2.41) and (2.42) with l taken as the distance from the rear wheel center to the center of the bogey free-pivot joint.

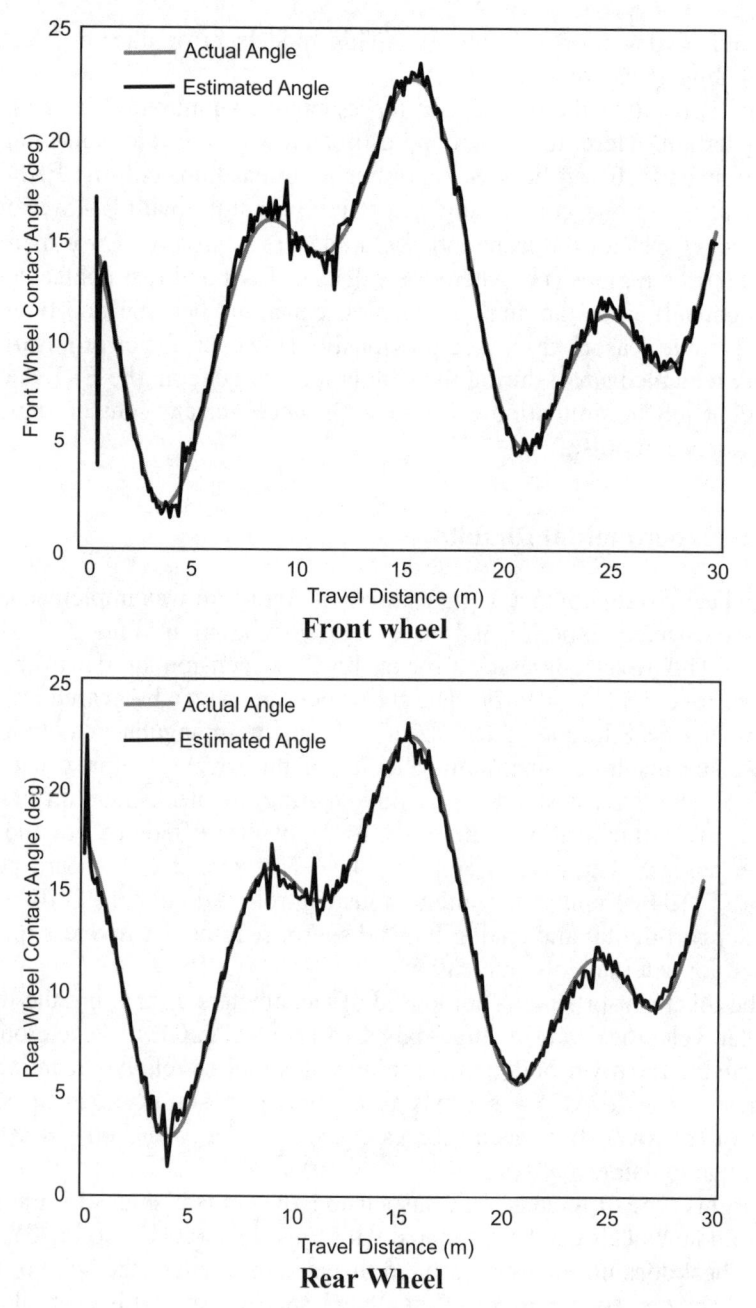

Fig. 2.16. Simulated wheel-terrain contact angles and estimates for front and rear wheels

2.6 Results: Wheel-Terrain Contact Angle Identification 47

Fig. 2.17. Experimental rover testbed

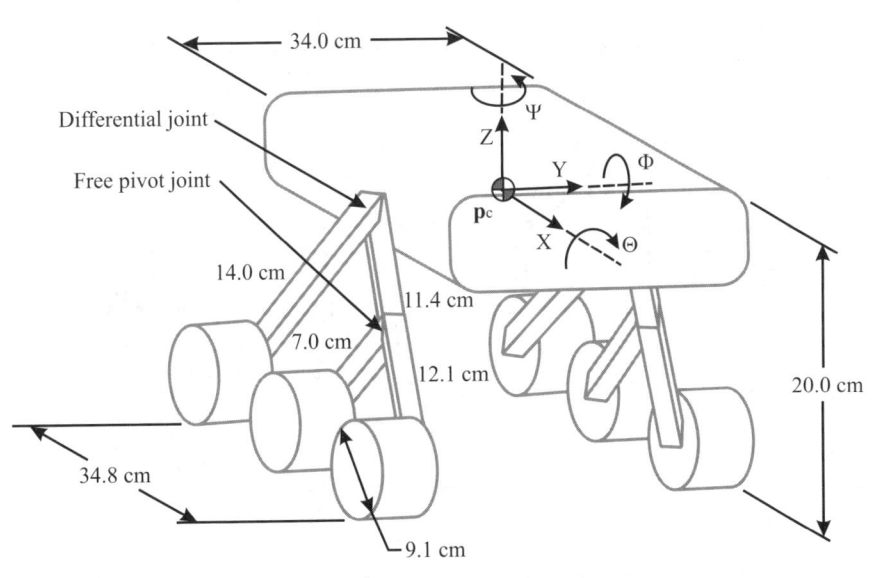

Fig. 2.18. Experimental rover testbed kinematic description

Measurement noise covariances for the EKF were estimated off-line by analyzing the standard deviation of sensor readings for the wheel tachometers and joint potentiometers during a trial motion. Process noise covariances were estimated based on the expected terrain variation.

Two experiments were performed. In the first experiment the rover was driven over a rock of approximately one-half wheel diameter in size. The wheel-terrain contact angles were estimated using the EKF framework described above. Results of this experiment can be seen in Fig. 2.19. The wheel-terrain contact angles increase as the wheel ascends the rock, decrease to zero atop the rock, and become negative as the wheel descends the rock. As expected, the contact angle estimates are similar for all three wheels.

In the second experiment the rover was driven from flat terrain up a 20° incline, and the front and middle wheel-terrain contact angles were estimated. Results of this experiment can be seen in Figure 2.20. The RMS error of the front wheel contact angle estimate is 2.21° while the middle wheel RMS error is 1.84°. Average computation time for both experiments was 0.6 msec/cycle, which is reasonable for on-board implementation. Thus it can be concluded that wheel-terrain contact angles can be accurately estimated by a rover with noisy sensors and limited computational resources.

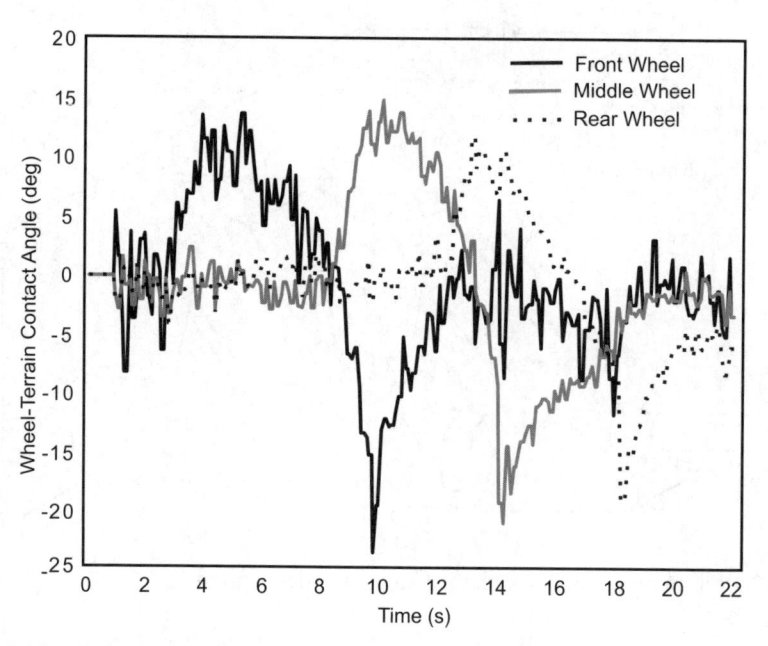

Fig. 2.19. Estimated wheel-terrain contact angle of front, middle and rear right-side wheels of experimental rover traversing a rock

2.6 Results: Wheel-Terrain Contact Angle Identification 49

Front Wheel

Middle Wheel

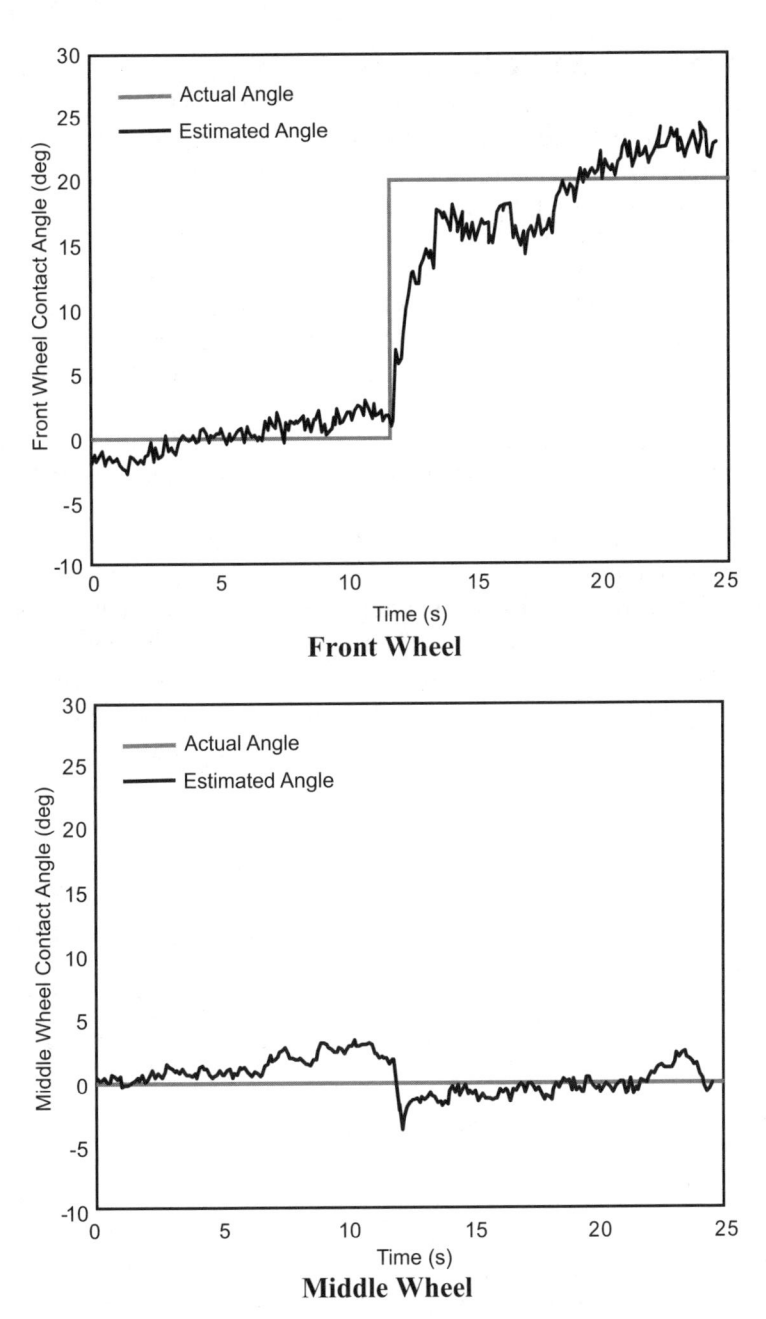

Fig. 2.20. Estimated and actual wheel-terrain contact angles for front and middle wheels of experimental rover traversing 20° incline

2.7 Summary and Conclusions

Terrain parameters and wheel-terrain contact angles are important elements of rough terrain robot models. Knowledge of terrain parameters and wheel-terrain contact angles allows accurate traversability prediction, and is useful for motion planning and control.

An efficient on-line terrain parameter estimation algorithm has been presented. The estimation method is based on simplified forms of classical terramechanics equations. A linear least-squares estimator was used to estimate cohesion and internal friction angle in real time. Simulation and experimental results have shown that the method can estimate parameters of three different terrain types with good accuracy despite noise, using limited computation. This method could be used for planetary rovers for on-line terrain analysis.

A wheel-terrain contact angle estimation algorithm based on rigid-body kinematic equations was also presented. The algorithm utilizes an extended Kalman filter to fuse on-board sensor signals. Simulation and experimental results for a six-wheeled rover have shown that the algorithm can accurately estimate wheel-terrain contact angles in rough terrain.

Chapter 3
Rough Terrain Motion Planning

3.1 Introduction

Motion planning algorithms for mobile robots in rough terrain must explicitly consider the effects of robot-terrain interaction. This differs from planning for systems operating on level, smooth surfaces, where terrain effects can usually be safely ignored. In rough terrain, it is crucial that a robot be able to assess mobility hazards posed by sloped, loose, and uneven terrain, in order to formulate a safe, robust motion plan.

Rough terrain motion planning algorithms must also consider real-world implementation issues. Specifically, algorithms must be efficient enough to be executed on a system with limited on-board computational resources, such as a planetary exploration rover. Also, rough terrain planning algorithms must consider the uncertainty inherent in terrain sensing systems such as rangefinders. Finally, planning algorithms must account for the fact that the mobile robot cannot precisely track a planned path due to path-following errors.

In this chapter two distinct rough terrain motion planning algorithms are presented. The first algorithm plans the route (i.e. the pointwise position and heading) of a mobile robot through uneven terrain. The purpose of this algorithm is to allow a mobile robot to safely move from its current position to a desired position while optimizing some aspect of system performance (such as power consumption). The second algorithm plans the suspension configuration of an actively articulated suspension robot as it travels through uneven terrain. This algorithm applies to mobile robots that possess actively controlled suspensions that can reposition their center of mass to improve mobility in rough terrain. Both algorithms can be considered motion planning algorithms, despite the fact that the first algorithm's solution lies in the space of the robot's position and heading and the second algorithm's solution lies in a space of the robot's suspension degrees-of-freedom. Both algorithm also consider important terrain effects.

The rough terrain motion planning technique presented in Sect. 3.2 considers terrain characteristics and relevant real-world implementation issues. Simulation results for this algorithm are presented in Sect. 3.3. It is shown that that the method is able to effectively plan safe routes through rough terrain. A configuration planning method for robots with actively articulated suspensions is presented in Sect. 3.4. Simulation and experimental results are shown in Sect. 3.5. It is shown that the algorithm can greatly improve robot stability in rough terrain.

3.2 Rough Terrain Motion Planning

In many rough terrain applications, mobile robots are required to autonomously plan a path from their current position to a distant goal location while avoiding physical obstacles and untraversable terrain regions. For example, a planetary exploration rover might be required to traverse rugged, boulder-strewn terrain to reach a scientifically interesting site in a steeply-sloped gully, with limited human intervention.

Motion planning in rough terrain differs from planning in well-structured environments in several respects. First, terrain data cannot be assumed to be perfectly known, due to errors in range sensing techniques and sensor miscalibration. Second, the planned path may not be accurately followed by the robot due to significant path-following and localization errors. Third, in rough terrain, the concept of an obstacle is not clearly defined, as it depends on an understanding of the terrain and the mobility characteristics of the robot. For example, smooth, flat terrain might be considered an obstacle if its traction characteristics are poor. Conversely, rugged terrain might be traversable for robots with well-designed suspensions. Finally, planetary exploration systems will generally have limited computational resources to devote to path planning.

In this section a motion planning method is presented that considers range data and robot localization uncertainty, and utilizes detailed physical models of the robot and its environment to assess path traversability. The input to the planning method is a terrain elevation map, such as would be obtainable from a laser rangefinder or stereo camera pair. The assumed range of this data is approximately 5-10 robot lengths.

The method is composed of two steps, as shown in Fig. 3.1. The first step is a rapid search through the terrain map for a candidate path, and is described in Sect. 3.2.1. The rapid path search uses a measure of local terrain roughness and a classical A^* graph search algorithm to quickly find a reasonable path through the range map from the robot start position to the goal position (Nilsson 1980). Terrain roughness is defined with respect to robot physical parameters such as wheel diameter.

The second step is a rigorous evaluation of the proposed path using robot and terrain physical models, and is described in Sect. 3.2.2. Uncertainty in terrain measurement and robot localization are considered. If the model-based evaluation determines that the robot could be subject to failure along the proposed path, the $A*$ cost function is increased at the potential failure location, and the path is replanned to avoid the hazard.

Simulation results in Sect. 3.3 show that the proposed method can efficiently and safely plan routes through rough terrain. The method is compared to a "conventional" motion planning method, and it is shown that the use of physical models of the robot and terrain allows the rough terrain planner to avoid hazards that the conventional planning methods cannot detect.

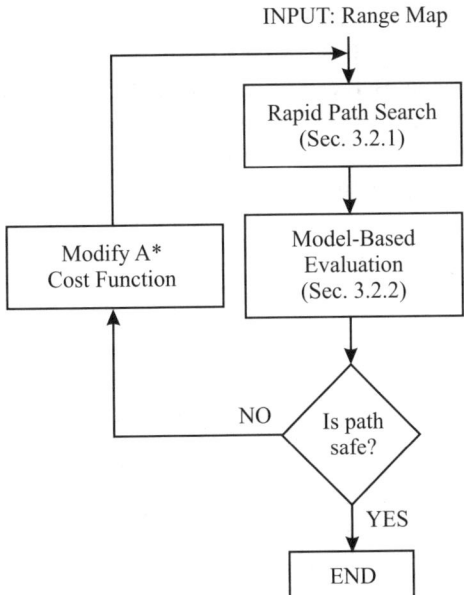

Fig. 3.1. High-level flow chart of the planning algorithm

3.2.1 Step One: Rapid Path Search

The purpose of the rapid path search is to quickly find a direct, reasonable path from the current robot position to the goal position. In the interest of reducing computation time for on-board implementation, the rapid search is not a globally optimal search. The input to the search is a 3D terrain range map, such as would be obtainable from a laser rangefinder or stereo camera pair. The terrain is represented as a map of elevations z associated with a grid in (x,y), as illustrated in Fig. 3.2.

An A^* algorithm is used to rapidly find a path through the terrain grid from the current robot position to the goal position (Nilsson 1980). The A^* algorithm is a graph-search technique, and is attractive due to its high speed for relatively small graphs. In this work, the assumed graph size is approximately 10^4 cells. The A^* algorithm computes a path based on a user-defined performance index, Φ.

Performance Index Definition

The performance index is formulated as a function of three variables. Terrain roughness, r, is considered since it is directly related to traversability and robot safety (Bekker 1969). Robot turning action, t, is considered, since in rough terrain excessive turning may not be desirable or possible, especially for skid-steered robots. Path length, l, is considered in order to minimize energy expenditure. Since the performance index must be evaluated a large number of times, it should be mathematically simple to speed computation.

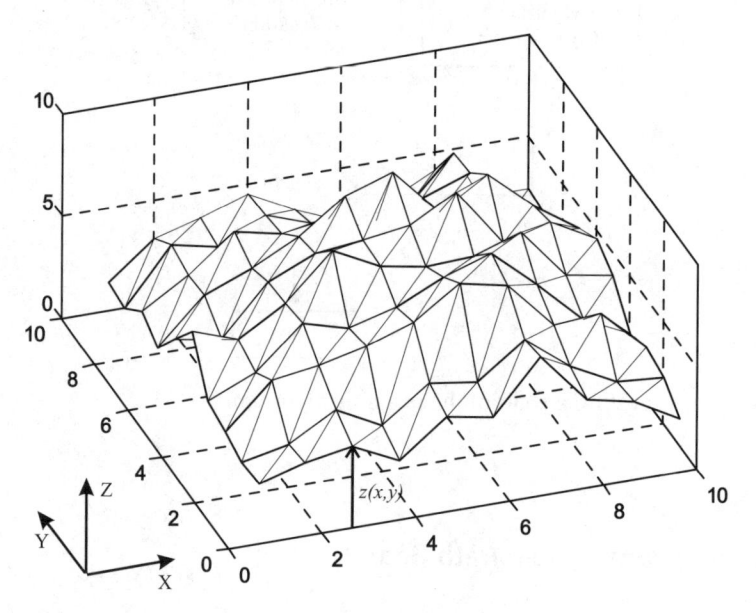

Fig. 3.2. Example of terrain data input to rapid path search

Terrain roughness can be defined in numerous ways. Here a roughness definition is proposed based on the variance of the terrain elevation over a robot-sized patch. Consider a robot centered at a point *(x,y)* and oriented in the direction of motion along a proposed path. Let $\Re$ be the set of terrain elevation points inside the convex hull defined by the wheel-terrain contact points of the robot on flat ground, as shown in Fig. 3.3. The terrain roughness *r* at *(x,y)* is defined as the square root of the variance of all elevation points in $\Re$:

$$r(x, y) = \frac{\sqrt{\text{var}(z(\Re))}}{d} \quad (3.1)$$

where *d* is the robot wheel diameter and is included to non-dimensionalize and scale the metric to the robot size. Note that it is important to include "interior" terrain points (i.e. inside the robot footprint) in order to provide an estimate of potential robot hang-up failure. Hang-up failure occurs when the robot body becomes lodged atop an obstacle, causing loss of wheel traction and robot entrapment.

Terrain areas with large roughness relative to the robot wheel diameter are likely untraversable. It is desirable to assign high cost to these regions. To accomplish this a modified roughness measure *r'* is defined, as:

$$r'(x, y) = \left(\frac{\sqrt{\text{var}(z(\Re))}}{d}\right)^{\alpha_1} \quad (3.2)$$

where α_1 is a constant greater than one. This has the effect of increasingly penalizing terrain regions that are clearly untraversable, while maintaining a continuous (as opposed to binary) obstacle representation.

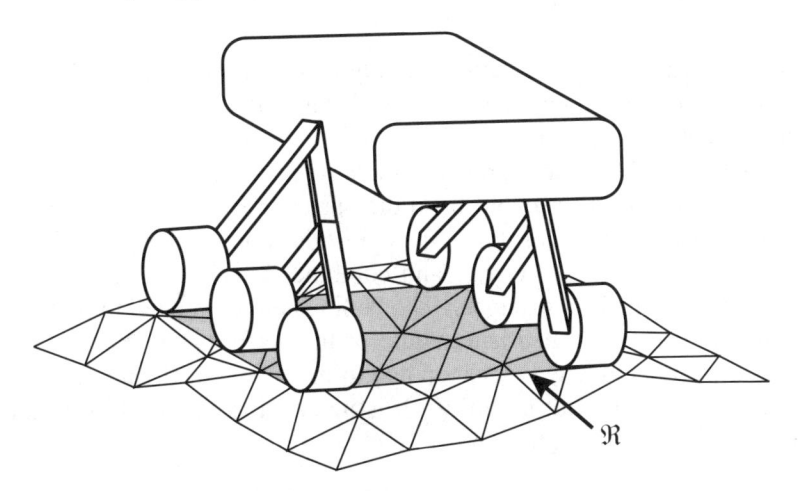

Fig. 3.3. Terrain roughness region definition

The cost associated with robot turning, t, is also defined as a function of terrain roughness. Turning in highly rough terrain is difficult since most robot suspensions are designed to surmount obstacles in their axial (i.e. forward) direction. Turning in rough terrain creates forces that increase the likelihood of rollover. Also, robots that are skid-steered cannot turn in rough terrain due to terrain-induced lateral forces acting on the wheels. A cost is therefore assigned to turning that is proportional to terrain roughness, at points where the proposed path changes heading. The turning cost function t is defined as:

$$t(x, y) = \left(\frac{\sqrt{\text{var}(z(\Re'))}}{d}\right)^{\alpha_2} \quad (3.3)$$

where α_2 is a constant greater than one and $\Re'$ is defined as the set of points inside a circle centered at (x,y) with radius equal to the distance from the robot center to the most distant wheel-terrain contact point. Thus, $\Re'$ can be viewed as a superset of $\Re$ augmented to include the area swept during turning. The wheel diameter is again utilized for scaling and non-dimensionalizing the metric.

The cost, l, associated with path length is simply taken as the length L of the candidate path divided by a robot wheel diameter:

$$l(x, y) = \frac{L}{d}. \quad (3.4)$$

The performance index Φ is formed as a weighted sum of the metrics considering terrain unevenness, turning, and path length, and a penalty term p:

$$\Phi = k_1 r' + k_2 t + k_3 l + p \quad (3.5)$$

with constants k_1, k_2, and k_3 chosen to adjust the relative values of r', t, and l to address mission-specific constraints. The penalty term p is zero for all points except those identified as hazardous by the model-based evaluation, described below. For those points it is a large positive number.

A least-cost path P is found from the current robot position to the goal position using the A^* algorithm. The path P is composed of n neighboring (x,y) pairs, $P = \{x_1\ y_1,...,x_n\ y_n\}$, and combines terrain smoothness, minimum turning, and short distance. The path is not guaranteed to be hazard-free, due to the simple heuristic nature of the performance index. Again, the performance index was selected due to its ease of evaluation and intuitive relationship to robot mobility. It is intended to rapidly lead to a reasonable path through the terrain.

3.2.2 Step Two: Model-Based Evaluation

The path generated by the rapid search is not guaranteed to be hazard-free. A detailed model-based evaluation of the proposed path is therefore required to ensure robot safety.

To perform a model-based evaluation, the robot configuration is first computed at each (x,y) pair in path P through a kinematic analysis, as described in Sect. 2.2.1. The robot configuration associated with each point can then be analyzed for static stability, kinematic validity, and terrain traversability. These analyses are described below.

Static Stability Analysis

Given a robot configuration, static stability can be computed in a manner similar to that proposed by (Papadopoulos and Rey 1996). Again, a static analysis is appropriate due to the low speeds of these vehicles. In this approach, the robot's m wheel-terrain contact points $\mathbf{p}_i$, $i=\{1,...,m\}$ are numbered in ascending order in a clockwise manner when viewed from above, as shown in Fig. 3.4. Note that all vectors in this analysis are expressed in the inertial frame *{XYZ}*. The lines joining the wheel-terrain contact points are referred to as tipover axes and denoted $\mathbf{a}_i$, where the ith tipover axis is given by:

$$\mathbf{a}_i = \mathbf{p}_{i+1} - \mathbf{p}_i, \quad i = \{1,..., m-1\}, \tag{3.6}$$

$$\mathbf{a}_m = \mathbf{p}_1 - \mathbf{p}_m. \tag{3.7}$$

A vehicle with m wheels or feet in contact with the terrain has in general m tipover axes. Tipover axis normals $\mathbf{l}_i$ that intersect the center of mass can be described as:

$$\mathbf{l}_i = \left(1 - \hat{\mathbf{a}}_i \hat{\mathbf{a}}_i^T\right)\mathbf{p}_{i+1} \tag{3.8}$$

where $\hat{\mathbf{a}} = \mathbf{a}/\|\mathbf{a}\|$.

Static stability angles can be computed for each tipover axis as the angle between the gravitational force vector $\mathbf{f}_g$ and the axis normal $\mathbf{l}_i$:

$$\eta_i = \sigma_i \cos^{-1}\left(\hat{\mathbf{f}}_g \cdot \hat{\mathbf{l}}_i\right), \quad i = \{1,...,m\} \tag{3.9}$$

with

$$\sigma_i = \begin{cases} +1, & \left(\hat{\mathbf{l}}_i \times \hat{\mathbf{f}}_g\right) \cdot \hat{\mathbf{a}}_i < 0 \\ -1, & \text{otherwise} \end{cases}. \tag{3.10}$$

58 Chapter 3 Rough Terrain Motion Planning

The overall vehicle stability angle is defined as the minimum of the m stability angles:

$$\alpha = \min(\eta_i), \quad i = \{1,\ldots,m\}. \tag{3.11}$$

When $\alpha < 0$ a tipover instability is occurring. Thus, a point in P is deemed a failure point if $\alpha < \alpha_{min}$, where α_{min} is a positive constant. This constant is user-defined, and can be viewed as a safety margin. For points identified as failure points, the penalty term p in Eq. (3.5) is increased to a large number.

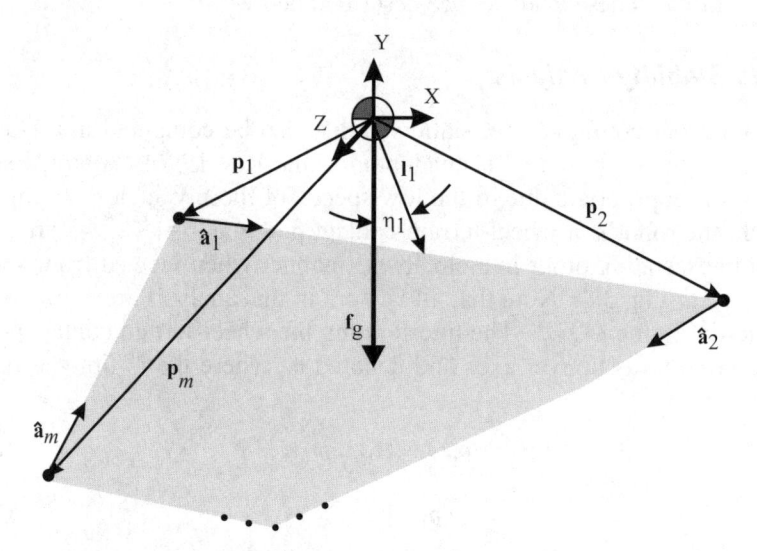

Fig. 3.4. Stability definition diagram

Kinematic Validity Analysis

A robot configuration is kinematically valid if no joint-limit or interference constraints are violated. To verify this, at each configuration in P the robot joint values are required to satisfy a set of inequality constraints:

$$\theta_i^{min} < \theta_i < \theta_i^{max}, \quad i = \{1,\ldots,q\} \tag{3.12}$$

where θ_i^{min} and θ_i^{max} can be functions of the robot configuration, and q is the number of robot suspension joints. If the inequality constraint is violated for any joint, the robot is unable to conform to the terrain. Physically, this implies that a robot wheel will lose contact with the terrain, which can lead to loss of traction. In this case the point is deemed a failure point and the penalty term p in Eq. (3.5) is increased to a large number.

Terrain Traversability Analysis

At each point in path P a quasi-static force analysis is performed to determine if the robot can exert enough force to move in the desired direction, as described in Sect. 2.2.2. The input force $\mathbf{f}_s$ to the force analysis is a small positive constant in the direction of robot motion along the path. Thus, if a solution to the force analysis exists, the robot can generate enough force to cause motion in the desired direction. If the analysis predicts that the robot cannot move in the desired direction, the point is deemed a failure point and the penalty term p in Eq. (3.5) is increased to a large number.

If at any point the stability, kinematic validity, or terrain traversability constraints are violated, the penalty term p is increased to a large number at these points. If no stability, kinematic validity, or terrain traversability constraints are violated, the proposed path P is deemed safe. Note that the path P represents an idealized motion plan that does not consider the effects of uncertainty and path following error. These factors are discussed next.

3.2.3 Uncertainty in Rough Terrain Motion Planning

There are numerous sources of uncertainty associated with rough terrain motion planning. Uncertainty primarily lies in terrain elevation measurement and robot path following error. The rapid path search and model-based evaluation presented above implicitly assume that terrain elevation data is error-free and the robot can follow a desired path with perfect accuracy. In practice, terrain elevation data contains error, and robots will deviate from a desired path, largely due to localization uncertainty. If these effects are not explicitly considered in the planning method, the robot may attempt to traverse a terrain region that is untraversable, which can lead to entrapment or hang-up failure. In this section models of uncertainty sources are presented.

Terrain Measurement Uncertainty

Terrain elevation measurement systems such as laser rangefinders and stereo-camera range sensors are subject to significant error (Hebert and Krotkov 1992; Matthies and Grandjean 1994; Matthies et al. 1995). This error can be decomposed into a random, noise-based component, and a systematic component due to sensor bias and miscalibration (Matthies and Grandjean 1994).

The random component of sensor error is dominated by noise. For stereo camera ranging systems, this noise has been shown to be approximately Gaussian in nature, and a quadratic function of range:

$$\sigma_{Z_r} = \sigma_r Z^2 \tag{3.13}$$

where Z is the distance from the sensor to the terrain point of interest, and σ_r is the standard deviation of the sensor noise. Note that σ_r can be characterized off-line (Matthies and Grandjean 1994).

The systematic component of sensor error is dominated by miscalibration. It has been shown to be an approximately linear function of range (Matthies et al. 1995):

$$\sigma_{Z_s} = K_s Z \tag{3.14}$$

where K_s is a constant that can be identified off-line through standard techniques. This error can be physically interpreted as the angular displacement of the true sensor location from the assumed location.

Combining Eq. (3.13) and (3.14), uncertainty in terrain measurement can be modeled as the sum of a random and systematic component, as:

$$\sigma_Z^2 = (K_s Z)^2 + (\sigma_r Z)^2. \tag{3.15}$$

Robot Localization Uncertainty

Localization refers to estimation of the robot's position and orientation with respect to a fixed reference frame (Borenstein et al. 1996). The most common method of localization is dead reckoning based on wheel odometry information. Recent localization methods combine wheel odometry with inertial and visual data to improve accuracy (Olson et al. 2001). In general there is error associated with this type of localization due to wheel slip, sensor noise, and (possibly) terrain unevenness (Matthies et al. 1995). This error causes a robot to deviate from a planned path due to incorrect estimation of its true location.

Localization error has been studied for general mobile robot systems and for specific planetary exploration rovers (Matthies et al. 1995; Volpe 1999). A linear relationship has been observed in the error in both the estimated distance traveled and estimated change in heading during turns:

$$\sigma_L^{pos} = K_{pos} d, \tag{3.16}$$

$$\sigma_L^{head} = K_{head}\Delta \qquad (3.17)$$

where d is the linear distance traveled, Δ is the angular change in heading, and K_{pos} and K_{head} are constants. These constants can be identified off-line via experimental analysis.

3.2.4. Incorporating Uncertainty in the Rapid Path Search

Uncertainty is considered in both the rapid path search and model-based evaluation. In this section, methods for incorporating the terrain measurement and localization uncertainty models into the rapid path search are described.

Terrain Measurement Uncertainty

The rapid search relies on terrain unevenness estimates to generate candidate paths. Thus it is important to consider terrain measurement uncertainty in the rapid search.

Terrain measurement uncertainty causes misrepresentation of the true position of terrain features. This is a threat to robot safety since erroneous range sensor data could lead to motion plans that are physically untraversable. Of particular practical importance is misrepresentation of the position (as opposed to elevation) of uneven terrain regions, since this might lead to motion plans that intersect with obstacles (such as large boulders) or untraversable regions (such as steep slopes). Thus it is important to consider the effects of terrain measurement uncertainty in the algorithm. To do this the terrain range map is pre-filtered with a two-dimensional Gaussian filter of the form:

$$z'(x,y) = \frac{e^{-\frac{x_r^2 + y_r^2}{2\sigma_Z^2}}}{\sigma_Z \sqrt{2\pi}} \qquad (3.18)$$

where the standard deviation σ_Z of the filter is defined by the terrain measurement uncertainty model described by Eq. (3.15) and x_r and y_r refer to the x and y distance, respectively, from the robot to the terrain point.

The filter has the effect of "blurring" terrain data, as illustrated in Fig. 3.5. The filtered range map possesses an enlarged area of uneven terrain compared to the unfiltered map. Thus the rapid search will impute a "safety margin" to the proposed path by planning a route that lies further from the potential obstacle.

Localization uncertainty is not explicitly considered in the rapid path search. It is considered in the model-based evaluation, discussed below.

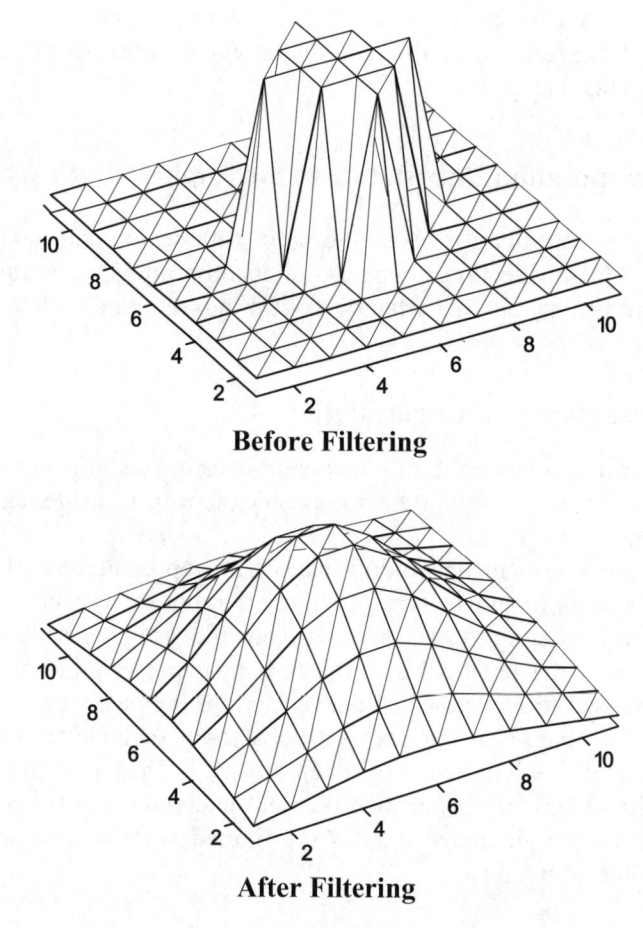

Before Filtering

After Filtering

Fig. 3.5. Example of obstacle elevation before and after Gaussian filter

3.2.5. Incorporating Uncertainty in the Model-Based Evaluation

The model-based evaluation verifies robot static stability, kinematic validity, and terrain traversability at discrete points along the proposed path. The model-based evaluation must account for imperfect knowledge of the terrain, and the fact that the robot may not accurately follow the proposed path due to localization error.

Terrain Measurement Uncertainty

Uncertainty in terrain measurement affects the model-based evaluation by creating ambiguity in the true robot configuration at a given terrain point. This affects computation of static stability, kinematic validity, and terrain traversability. Due to the high cost associated with robot failure, a conservative approach to considering terrain measurement uncertainty is adopted. In this approach, the set of potential worst-case robot configurations at a given terrain point are computed and analyzed.

Potential worst-case configurations generally exist at possible terrain point location boundaries. Here we consider variation in elevation only, to speed computation. The elevation boundaries of a given terrain point z_n can be computed from Eq (3.15) as:

$$z_n^+ = z_n + \sigma_Z, \quad z_n^- = z_n - \sigma_Z. \tag{3.19}$$

For an n-wheeled robot, $m = 2^n$ potential worst-case robot configurations $Q' = \{Q_1,...,Q_m\}$ are computed by solving the inverse kinematic solution of the robot at each unique terrain point combination, as illustrated in Fig. 3.6. Although sophisticated techniques exist for dealing with uncertainty propagation, due to the relatively small number of potential configurations a brute-force analysis is appropriate (Latombe 1991; Page and Sanderson 1995; Zhang et al. 1998).

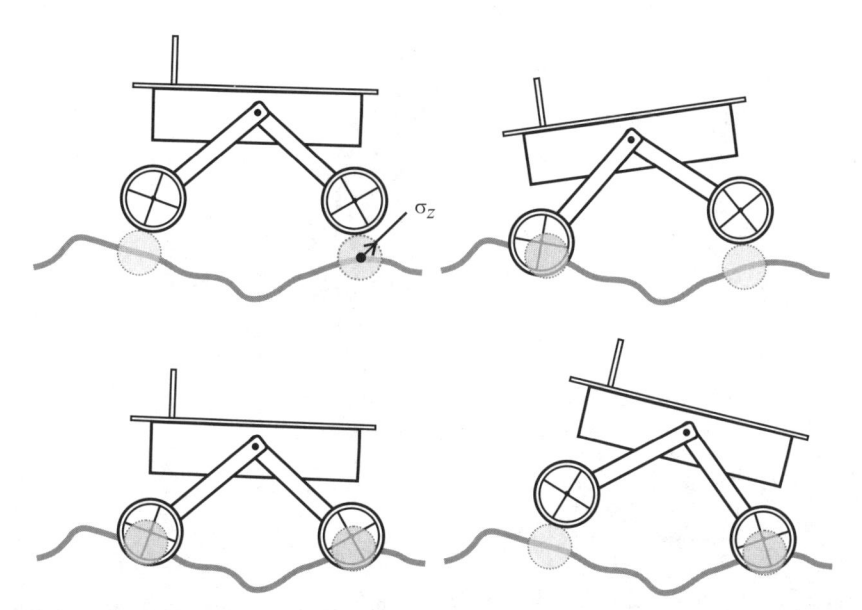

Fig. 3.6. Set of possible configurations of a planar robot due to terrain measurement uncertainty

Each configuration in Q' is examined for static stability, kinematic validity, and terrain traversability as described in Sect. 3.2.2. If any configuration is deemed unsafe, the point is deemed a failure point and the penalty term p in Eq. (3.5) is increased to a large number. The rapid path search is then called to re-plan the path.

Note that from Eq. (3.15), terrain measurement uncertainty increases with range distance. Clearly, at some finite distance the uncertainty will increase to a point where no kinematically valid robot configuration can be found. This places an upper bound on the range of the planning algorithm independent of the maximum sensing distance of the range sensor.

Robot Localization Uncertainty

Robot localization uncertainty affects the model-based evaluation by increasing the number of possible configurations a robot may experience during traversal of the proposed path. To account for this, a superset P' of P can be formed by including all points near P within a distance proportional to the robot localization uncertainty, as illustrated in Fig. 3.7.

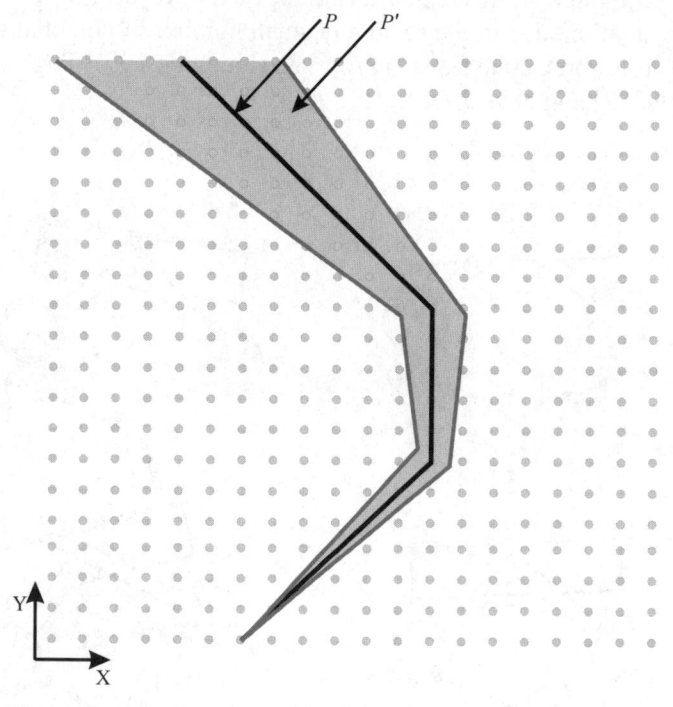

Fig. 3.7. Illustration of effect of localization uncertainty on model-based evaluation

All points in P' are examined for static stability, kinematic validity, and terrain traversability as described in Sect. 3.2.2. If any configurations associated with points in P' are deemed unsafe, the point is deemed a failure point and the penalty term p in Eq. (3.5) is increased to a large number. The rapid search is then called to re-plan the path. If the configurations in P' are deemed safe, the motion planning problem is considered solved and the algorithm ends.

3.3 Simulation Results—Rough Terrain Planning

Simulations were performed of the six-wheeled rover described in Sect. 2.6.2 traversing rough terrain. The purpose of the simulations was to compare the rough terrain motion planning method to a conventional planning method that does not utilize physical models, does not consider uncertainty, and treats obstacles in a binary manner. Three simulated terrains of increasing difficulty were generated by a pseudo-random terrain generator (Pickover 1995) (see Fig. 3.8).

Ten trials were performed for each terrain. Start and goal locations were randomly-generated, with a minimum straight-line spacing of 5 m. Each terrain map was a grid of 195 x 195 equally-spaced terrain points, with a uniform square spacing of 4 cm. This corresponded to a map approximately 8 m square. For the rough terrain planning method, the exponents α_1 and α_2 in Eqs. (3.2) and (3.3) were taken as 3. The weighting factors k_1, k_2, and k_2 of the performance index defined in Eq. (3.5) were 5.0, 0.1, and 0.1, respectively, and the value of p was 1×10^4. The planning method was simulated on a 300 MHz Pentium PC.

The rough terrain planning method was compared to a conventional planning algorithm that attempted to find the shortest path between the start and goal while avoiding obstacles. Here, an obstacle was defined by considering the maximum change in elevation between a terrain point and its eight nearest neighbors. If the maximum change in elevation was greater than 80% of the robot wheel diameter, the terrain point was considered an obstacle. The obstacle "cutoff" height was chosen heuristically based on knowledge of the robot system. Based on this definition, the terrain maps in Fig. 3.8 had obstacle densities of 0.05, 0.23, and 0.72 obstacles/robot area. Thus, the simulated terrains range from benign to difficult.

Benign Terrain

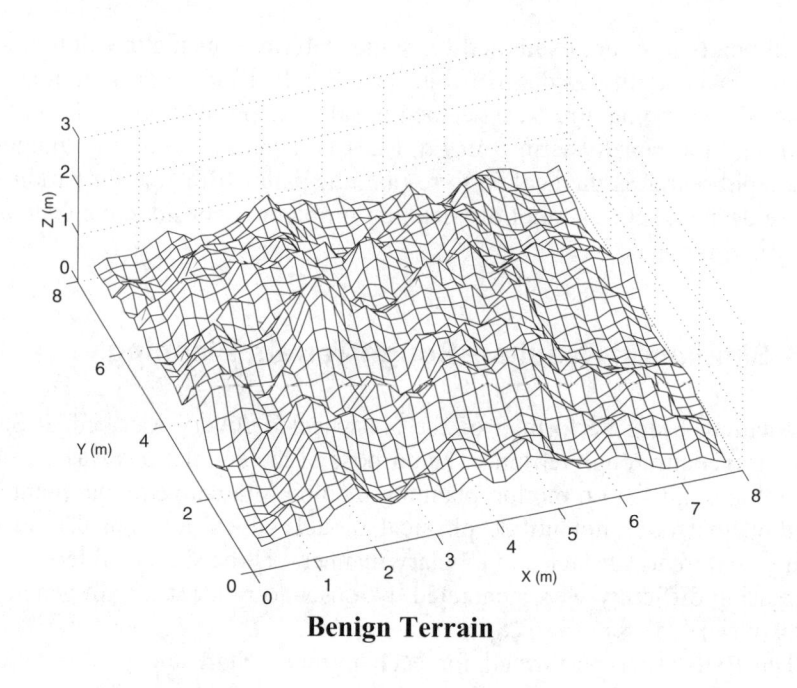

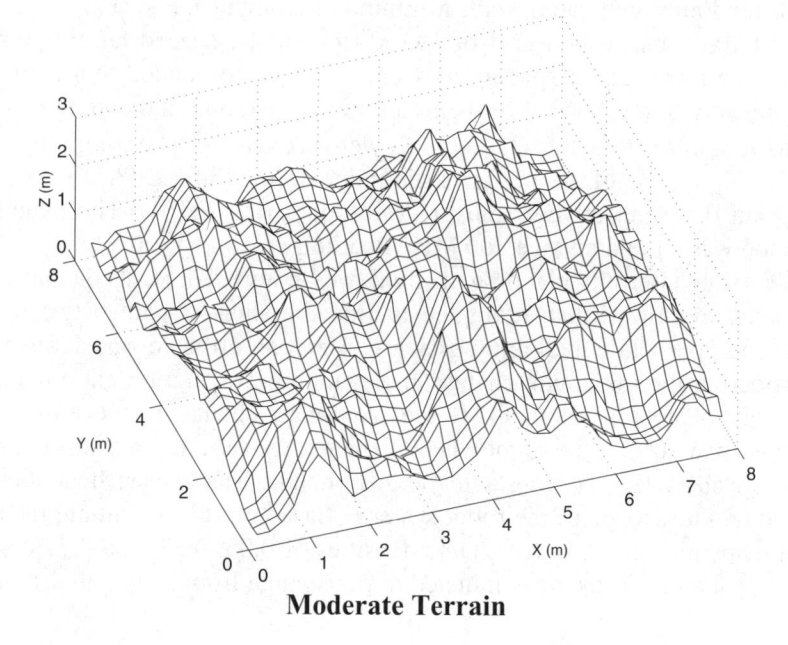

Moderate Terrain

Fig. 3.8. Simulated terrain elevation maps: Benign, moderate, and difficult

3.3 Simulation Results—Rough Terrain Planning

Difficult Terrain

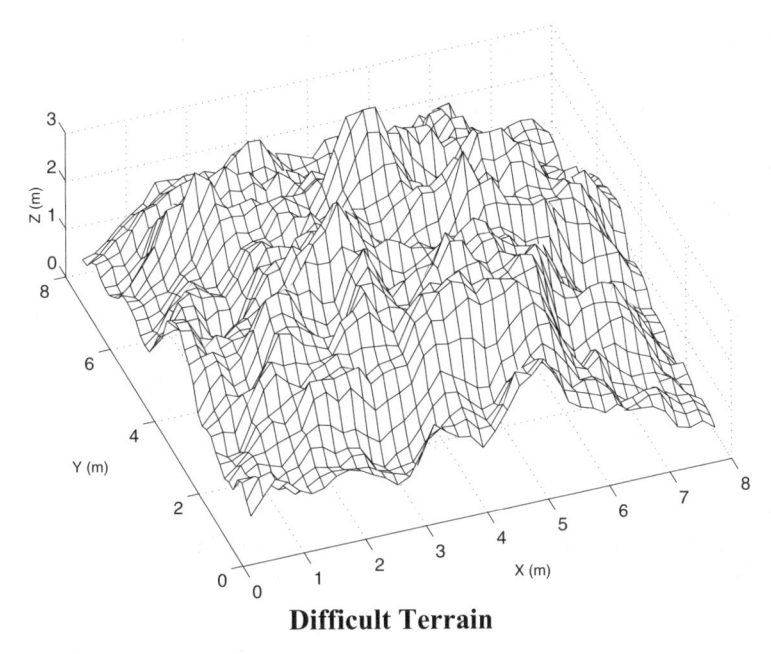

Fig. 3.8. (Continued)

Results of the simulation trials are summarized in Table 3.1. Rough terrain planning (RTP in the table) found safe paths that were on average 7.35% shorter than the binary planning method (BP in the table). This is due to the fact that the physical models employed by the rough terrain motion planning method allow it to plan paths through regions that a binary planning method would consider an obstacle. Since path length is related to power consumption, the rough terrain planning method is more power-efficient than a conventional planning method.

Table 3.1. Results of motion planning algorithm comparison

		Average Path Length (m)	Average Computation Time (s)	Number of Reachable Goals
Benign Terrain	RTP	6.02	35.5	10
	BP	6.44		10
Moderate Terrain	RTP	6.34	40.5	10
	BP	6.95		9
Difficult Terrain	RTP	6.95	118.1	9
	BP	7.48		6

Arguably more important than path length is the result that the rough terrain planning methods resulted in an increased number of reachable goals, compared to the binary planning method. For the highly difficult terrain shown in Fig. 3.8, the rough terrain planning method found 50% more safe paths than the binary planning method. This is significant for many applications such as planetary exploration, where terrain accessibility is of primary concern.

A representative simulation trial is presented in Fig. 3.9. Paths through the moderate terrain of Fig. 3.8 as found by the rough terrain planner and the binary planner are presented. It can be seen that the path generated by the rough terrain planner is more direct than the one generated by the binary planner. This is due to the fact that the binary planner viewed the crest at *(x,y)* = (2.75, 2.10) as an obstacle and avoided it. The rough terrain planner analyzed the crest and determined it was traversable, and thus proceeded across it.

A second representative simulation trial is shown in Fig. 3.10. Here the cutoff height for obstacle detection in the binary planner increased to 120% of the wheel diameter. This is an aggressive assumption about terrain traversability, which might be implemented to allow the robot to traverse highly rough terrain. Again, however, it is based on heuristic knowledge of the robot capabilities, and is not justified by rigorous model-based analysis. A route was planned through the difficult terrain shown in Fig. 3.8. This route was deemed safe by the binary planning algorithm. However, review of the path by the model-based analysis showed a kinematic validity failure at point *(x,y)* = (2.80, 3.00) and a terrain traversability failure at point *(x,y)* = (3.40, 3.40). The failure points were avoided in the rough terrain planning algorithm. This result illustrates the binary planning algorithm's lack of physical knowledge of the robot system leading to a potential failure situation.

From these results, it is clear that the rough terrain planning algorithm plans paths that are shorter and safer than a traditional planning method, and allows access to a larger percentage of the terrain. All of these factors are important considerations in autonomous planetary exploration.

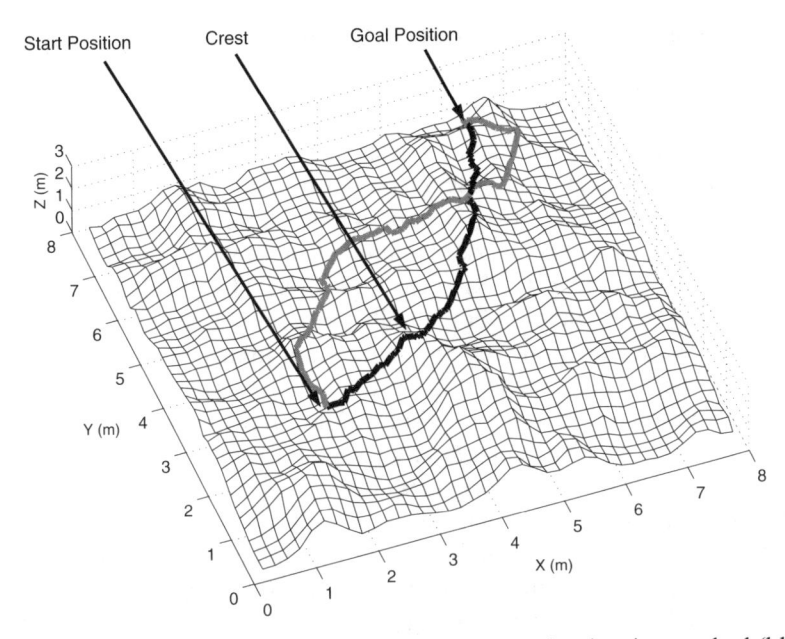

3.3 Simulation Results—Rough Terrain Planning 69

Fig. 3.9. Representative simulation trial for rough terrain planning method (black) and binary planning method (gray)

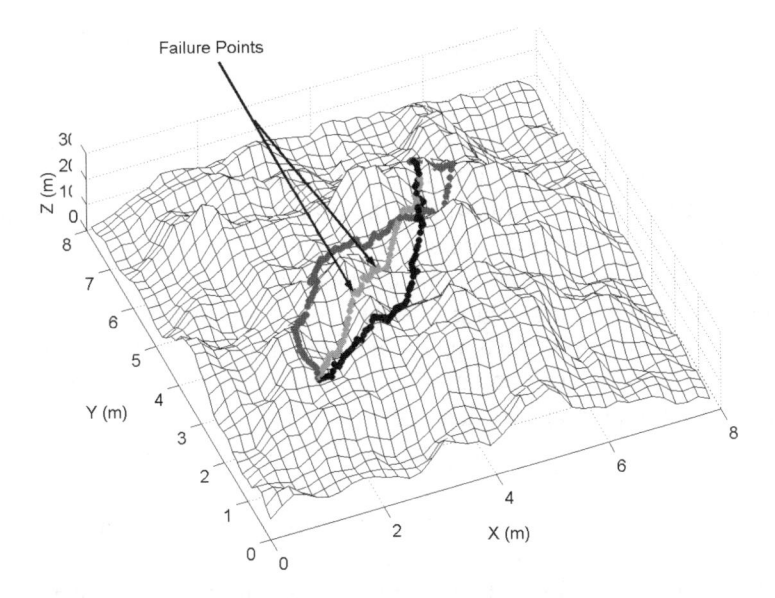

Fig. 3.10. Simulation trial for rough terrain planning (black), binary planning method with obstacle criteria of 80% of one wheel diameter (dark gray), and binary planning with obstacle criteria of 120% of one wheel diameter (light gray)

3.4 Rough Terrain Articulated Suspension Configuration Planning

Robots with actively articulated suspensions can improve rough terrain mobility by modifying their suspension configuration and thus repositioning their center of mass. In this section a method for stability-based articulated suspension configuration planning is presented, and demonstrated experimentally on the JPL SRR (see Fig. 1.4). The problem is stated in Sect. 3.4.1 and a general robot mobility analysis is performed in Sect. 3.4.2. Kinematic equations relating suspension joint variables to a vehicle stability measure are formulated in Sect. 3.4.3. The stability measure considers gravitational forces due to robot weight. It also considers forces due to manipulation, which are potentially large and could have a destabilizing effect on the vehicle. A performance index is defined based on the stability measure and a function that maintains adequate ground clearance, an important consideration in rough terrain. A rapid and computationally inexpensive conjugate-gradient optimization of the performance index is performed subject to vehicle kinematic constraints. Simulation and experimental results in Sect. 3.5 show that articulated suspension configuration planning can greatly improve robot stability in rough terrain.

3.4.1 Articulated Suspension Configuration Planning Problem Description

Consider a general n-link tree-structured wheeled mobile robot on uneven terrain, as shown in Fig. 3.11. The n links can form hybrid serial-parallel kinematic chains. It is assumed that the robot's l joints are active revolute or prismatic joints, and their values are denoted θ_i, $i = \{1,...,l\}$. It is also assumed that the wheels make point contact with the terrain, with m wheel-terrain contact points denoted $\mathbf{p}_j$, $j = \{1,...,m\}$. Vectors from the points $\mathbf{p}_j$ to the vehicle center of mass $\mathbf{p}_c$ are denoted $\mathbf{V}_i = [V_i^x \ V_i^y]^T$, $i = \{1,...,n\}$. The wheel-terrain contact angles at each point $\mathbf{p}_j$ are measured with respect to the horizontal axis and are denoted $\gamma_j, j = \{1,...,m\}$.

The goal of articulated suspension configuration planning is to improve mobility by modifying the suspension variables θ_i to optimize a user-specified performance index, Φ. This performance index can be selected to assess static stability, wheel traction, vehicle pose for optimal force application, ground clearance, or a combination of metrics, and is generally a function of the suspension and manipulation degrees of freedom. In this paper, optimization of static stability and ground clearance is addressed.

3.4 Rough Terrain Articulated Suspension Configuration Planning

Problem constraints take the form of joint limit and mechanical interference constraints. Constraints also arise from system mobility limitations. These constraints are discussed below.

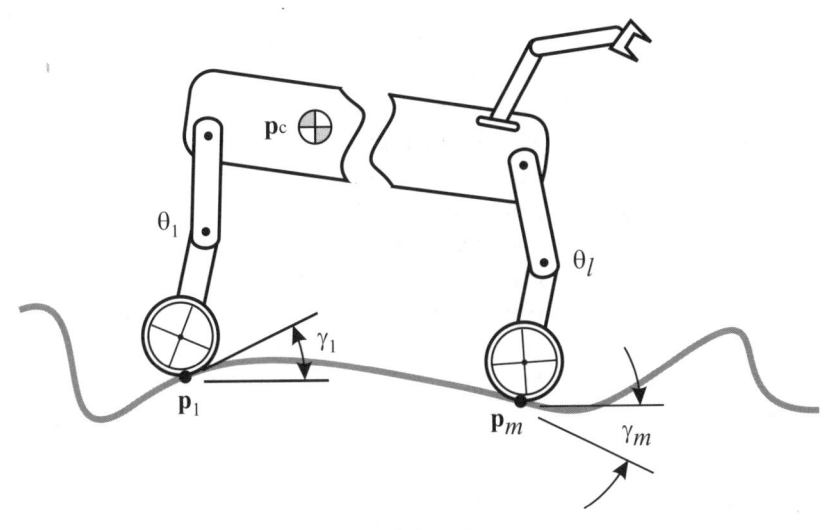

Fig. 3.11. A general tree-structured mobile robot

3.4.2 Mobility Analysis

The mobility of an articulated suspension robot can be analyzed using the Gruebler mobility criterion (Eckhardt 1989):

$$F = 6(l - j - 1) + \sum_{i=1}^{j} f_i \qquad (3.20)$$

where j is the number of joints, l is the number of links including the ground, and f_i is the number of constraints for each joint i. Some highly articulated robots may have mobility greater than or equal to one while stationary on the terrain (i.e. the robot has available self-motions). In these situations the terrain profile does not influence the configuration planning process. Thus, knowledge of robot kinematics alone is sufficient to pose the optimization problem.

Many articulated suspension robots, however, have a mobility less than or equal to zero. Thus, one or more wheel-terrain contact points p_j must move relative to the terrain during the configuration planning process, as in the examples shown in Figs. 1.4 and 1.5. Note that in such cases, wheel-terrain contacts must be treated as higher-order pairs during mobility analysis (Eckhardt 1989). In these situations, it is impossible to find a globally optimal solution for the suspension configuration without knowledge of the terrain profile. This is problematic, since the terrain profile is often not well known. However, the local wheel-terrain contact angles can be estimated, as described in Sect. 2.5.

The wheel-terrain contact angle γ_j describes the terrain profile in a local region about the point $\mathbf{p}_j$. An optimization problem can therefore be posed with the constraint that the robot suspension change results in only small displacements of the points $\mathbf{p}_j$ relative to the terrain (see Fig. 3.12). Here we assume that the terrain profile does not change significantly within a small region about the wheel-terrain contact points. Thus, a locally optimal solution for the suspension configuration can be found. Optimization constraints take the form of kinematic joint limit and interference constraints, and joint excursion limits that restrict the displacements of the points $\mathbf{p}_j$.

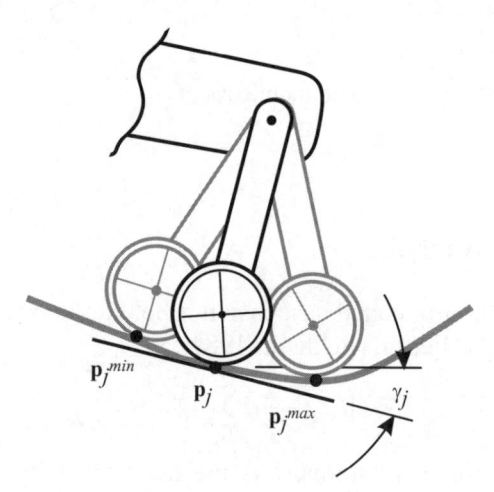

Fig. 3.12. Limited motion of wheel-terrain contact point $\mathbf{p}_j$

3.4.3 Articulated Suspension Configuration Planning for Enhanced Tipover Stability

Articulated suspension configuration planning can be used to improve criteria such as tipover stability or traction. In this section a method for enhancing static stability is described. Again, a static analysis is valid since planetary rovers travel at maximum speeds on the order of 10 cm/sec.

3.4 Rough Terrain Articulated Suspension Configuration Planning

In this work, vehicle stability is defined in the manner described in Section 3.2.2 (Papadopoulos and Rey 1996). In this approach, knowledge of the robot's kinematic configuration allows computation of a vehicle stability angle α (refer to Eqs. 3.6-3.11). When $\alpha < 0$ a tipover instability occurs. Measurements of wheel-terrain contact forces or articulation torques are not required to determine stability, as this is a kinematics-based stability analysis. The goal of articulated suspension configuration planning is therefore to maintain a large value of α.

In addition to traversing rough terrain, a robot may be required to manipulate its environment. Some manipulation tasks, such as sampling or coring, may require the application of large forces, which can destabilize the robot. During these tasks it would be desirable for the robot to optimize its suspension to maximize stability.

To account for manipulation forces in the stability computation, the applied force $\mathbf{f}_m$ is projected along a tipover axis as:

$$\mathbf{f}_i = \left(1 - \hat{\mathbf{a}}_i \hat{\mathbf{a}}_i^T\right)\left(\mathbf{f}_g + \mathbf{f}_m\right) \qquad (3.21)$$

with $\mathbf{f}_m$ expressed in an inertial frame. If there is a moment $\mathbf{n}_m$ associated with $\mathbf{f}_m$, the net force about a tipover axis is computed as:

$$\mathbf{f}_i = \left(1 - \hat{\mathbf{a}}_i \hat{\mathbf{a}}_i^T\right)\left(\mathbf{f}_g + \mathbf{f}_m\right) + \frac{\hat{\mathbf{l}}_i \times \left(\hat{\mathbf{a}}_i \hat{\mathbf{a}}_i^T\right)\mathbf{n}_m}{\|\mathbf{l}_i\|} \qquad (3.22)$$

The stability angle α is then computed from Eqs. (3.10-3.11) using the net force $\mathbf{f}_i$ in place of $\mathbf{f}_g$.

To optimize the robot suspension for maximum stability, a performance index Φ is defined based on the above stability measure. A function of the following form is proposed:

$$\Phi = \sum_{i=1}^{n} \left(\frac{K_i}{\eta_i} + K_{n+i}(\theta_i - \theta_i')^2 \right) \qquad (3.23)$$

where η_i are the stability angles defined in Eq. (3.9), θ_i' are the nominal values of the ith joint variables (i.e. the values of θ_i when the robot is at a user-specified configuration, such as on flat terrain), and K_i are constant weighting factors.

The first term of Φ tends to infinity as the stability at any tipover axis tends to zero. The second term penalizes deviation from a nominal configuration of the shoulder joints. This term is used to maintain adequate ground clearance and avoid obstacles that might cause hang-up, an important consideration in rough terrain. Note that an explicit guarantee of ground clearance would require analysis of 3-d terrain range data immediately in front of the vehicle. The proposed method is intended to be reactive in nature, and thus forward-looking range information is not considered. The constants K_i are selected to control the relative importance of vehicle stability and joint excursion.

The goal of the optimization is to minimize the performance index Φ subject to joint-limit, interference and possibly kinematic mobility constraints. Since Φ possesses a simple form for many systems, a rapid optimization technique such as the conjugate-gradient search can be employed (Arora 1989). The conjugate gradient algorithm minimizes a positive definite quadratic function. For a nonlinear function such as Eq. (3.23), the algorithm can be applied by interpreting the quadratic function as a second-order Taylor series approximation of the performance index Φ. This is a reasonable assumption for the kinematics-based performance index, which is composed of trigonometric functions.

Since this is a nonlinear optimization problem, local minima may exist in the search space. Practically, however, the existence of local minima is unlikely due to the small size of the search space. More complex and computationally expensive optimization methods which are robust to local minima can be employed, however they are unattractive since computation speed is important for planetary exploration rovers. The algorithm is summarized in Fig. 3.13.

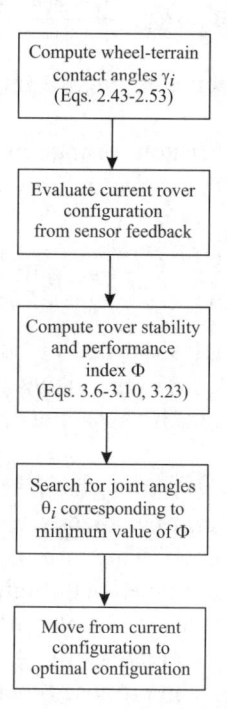

```
┌─────────────────────────┐
│ Compute wheel-terrain   │
│ contact angles γᵢ       │
│ (Eqs. 2.43-2.53)        │
└───────────┬─────────────┘
            ▼
┌─────────────────────────┐
│ Evaluate current rover  │
│ configuration           │
│ from sensor feedback    │
└───────────┬─────────────┘
            ▼
┌─────────────────────────┐
│ Compute rover stability │
│ and performance         │
│ index Φ                 │
│ (Eqs. 3.6-3.10, 3.23)   │
└───────────┬─────────────┘
            ▼
┌─────────────────────────┐
│ Search for joint angles │
│ θᵢ corresponding to     │
│ minimum value of Φ      │
└───────────┬─────────────┘
            ▼
┌─────────────────────────┐
│ Move from current       │
│ configuration to        │
│ optimal configuration   │
└─────────────────────────┘
```

Fig. 3.13. Algorithm summary

3.5 Results—Rough Terrain Articulated Suspension Configuration Planning

Simulations and experiments of the articulated suspension configuration planning algorithm were performed on the Jet Propulsion Laboratory Sample Return Rover traversing rough terrain. The SRR is a 7 kg, four-wheeled mobile robot with independently driven wheels and independently controlled shoulder joints, shown in Fig. 1.4 (Huntsberger et al. 1999). A 2.25 kg three d.o.f. manipulator is mounted at the front of the SRR. Controllable shoulder joints and manipulator allow the SRR to reposition its center of mass. The SRR is equipped with an inertial navigation system to measure body roll and pitch. Since the ground speed of the SRR is typically 6 cm/sec, dynamic forces do not have a large effect on system behavior, and thus static analysis is appropriate.

The optimization performance index used in the simulation and experiments was similar to Eq. (3.23) and considered the two shoulder angle joints θ_1 and θ_2 and the three manipulator degrees of freedom ψ_1, ψ_2, and ψ_3:

$$\Phi = \sum_{j=1}^{4} \frac{K_j}{\eta_j} + \sum_{i=1}^{2} K_{i+4} (\theta_i - \theta_i')^2 \qquad (3.24)$$

Note that the stability angles η_j are functions of the shoulder and the manipulator degrees of freedom (i.e. $\eta_j = \eta_j(\theta_1, \theta_2, \psi_1, \psi_2, \psi_3)$).

3.5.1 Simulation Results

The articulated suspension configuration planning algorithm was implemented in a planar simulation on steeply rolling terrain. The maximum pitch induced by the terrain was approximately 30°. The simulated pitch sensor was corrupted with white noise of standard deviation 3°. The rear and front wheel velocities, v_1 and v_2, were corrupted with white noise of standard deviation 0.5 cm/sec. This models error due to effects of wheel slip and tachometer noise.

Fig. 3.14 shows a representative simulation result. Vehicle stability margin as defined by Eq. (3.24) is shown for both articulated suspension and fixed suspension systems. The mean stability of the articulated suspension system was 37.1% greater than the fixed suspension system. The stability margin of the fixed suspension system reaches a minimum value of 1.1° (at travel distance $\approx$ 16.5 m), indicating that the system narrowly avoided tipover failure. The minimum stability margin of the articulated

76   Chapter 3   Rough Terrain Motion Planning

suspension system was 12.5°, a comfortable margin. This suggests that the articulated suspension configuration planning algorithm can substantially increase vehicle stability on uneven terrain.

3.5.2 Experimental Results

Numerous experiments were performed on the SRR in the JPL Planetary Robotics Laboratory and at an outdoor rough terrain test field, the Arroyo Seco in Altadena, California. The SRR was commanded to traverse a challenging path that threatened vehicle stability. For each trial the path was traversed first with the shoulder joints fixed, and then with the articulated suspension configuration planning algorithm activated. During these experiments, the SRR employed a state-machine control architecture, in which the vehicle traveled a small distance, stopped, then adjusted its shoulder angles based on the articulated suspension configuration planning algorithm. Fig. 3.15 shows the SRR traversing a steep incline with both fixed and articulated suspensions.

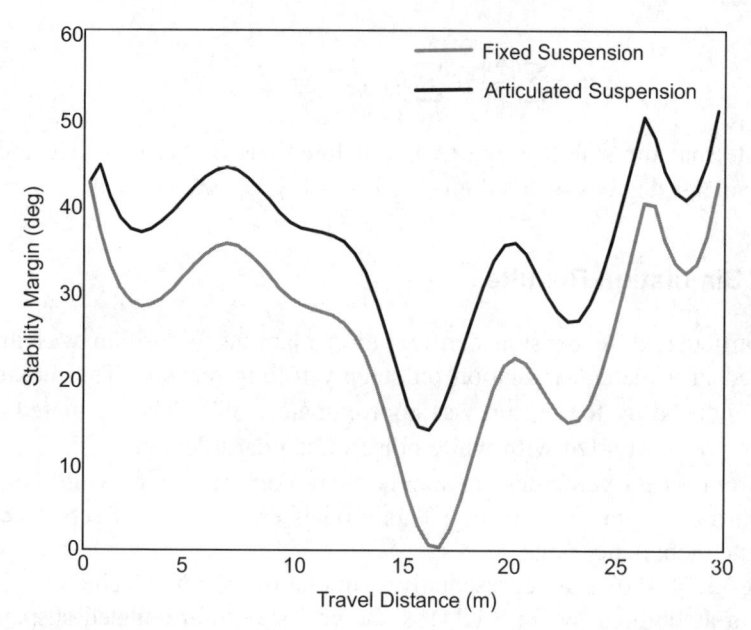

Fig. 3.14. Stability margin for robot with articulated suspension and fixed suspension during simulation

Fixed suspension   **Articulated suspension**

Fig. 3.15. SRR During rough terrain traverse

Results of a representative pair of trials are shown in Figs. 3.16 and 3.17. Fig. 3.16 shows the SRR shoulder joint angles during the traverses. Both left and right shoulder angles remain within the joint limits of ±45° of the initial values. Note that the fixed suspension shoulder angles vary slightly due to servo compliance.

Fig. 3.17 shows vehicle stability during the traverses. The average stability of the articulated suspension system was 48.1% greater than the fixed suspension system. The stability margin of the fixed suspension system reached dangerous minimum values of 2.1° and 2.5°. The minimum stability margin of the articulated suspension system was 15.0°. Clearly, articulated suspension configuration planning results in greatly improved stability in rough terrain.

Optimization was performed on-line with a 300MHz AMD K6 processor. Average processing time for a single constrained optimization computation was 40 μsec. Thus, the algorithm is feasible for on-board implementation.

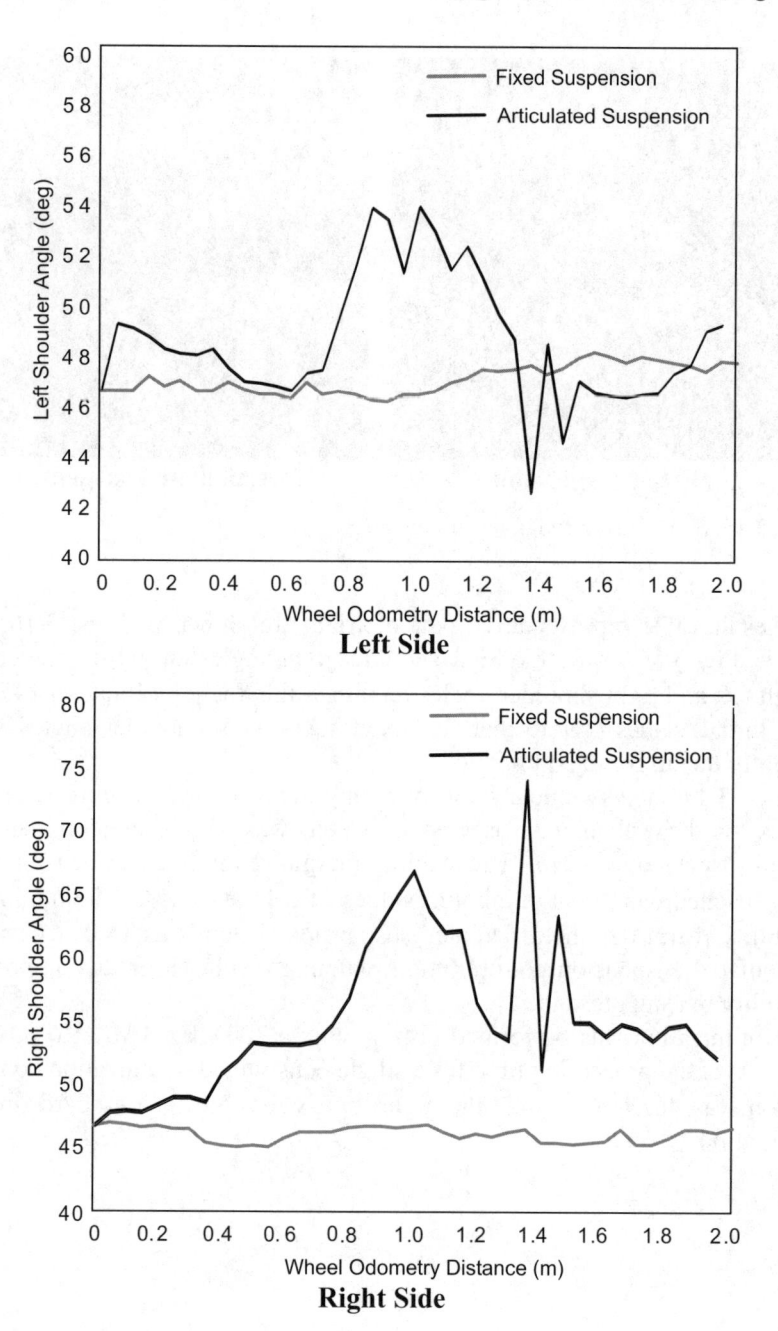

Fig. 3.16. SRR shoulder angles during rough terrain traverse for articulated suspension system and fixed suspension system

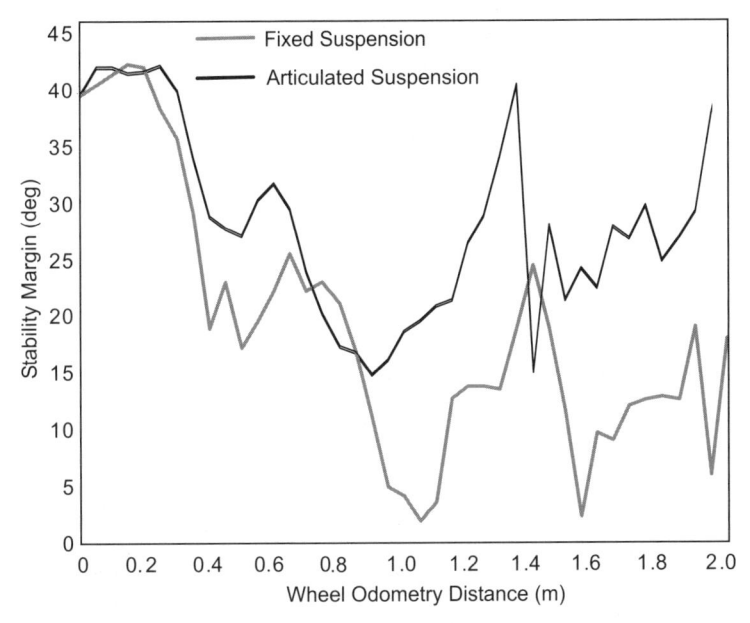

Fig. 3.17. SRR stability margin for articulated suspension system and fixed suspension system on uneven terrain

3.6 Summary and Conclusions

Rough terrain motion planning algorithms must explicitly consider the effects of terrain unevenness and traction characteristics, and real-world implementation issues such as computational efficiency and sensor uncertainty. In this chapter two distinct rough terrain motion planning algorithms have been presented.

The first algorithm plans the route of a mobile robot through uneven terrain. Uncertainty in terrain range data, terrain modeling, and robot path-following are considered. Simulation results demonstrates the effectiveness of the method.

The second algorithm plans the suspension configuration of an articulated suspension robot as it travels through uneven terrain. The method rapidly optimizes a performance index based on robot stability to determine an improved suspension configuration. Simulation and experimental results for the JPL SRR show that the articulated suspension configuration planning method yields greatly improved vehicle stability in rough terrain.

Chapter 4
Rough Terrain Control

4.1 Introduction

Effective control methods for mobile robots in rough terrain should maximize vehicle mobility despite changing terrain unevenness and traction characteristics. They should also minimize power consumption, since autonomous systems often possess limited on-board power storage capabilities. One means of improving control performance is to take advantage of the actuation redundancy present in most mobile robots. Actuation redundancy refers to the condition in which a mobile robot has more controllable drive actuators than is minimally required for locomotion. For example, a mobile robot with four driven wheels possesses actuation redundancy, since only two driven wheels are required to drive forward and turn via skid steering.

In this chapter a rough terrain control methodology is presented for robots with actuation redundancy. The fundamental control problem is posed in Sect. 4.2, and problem constraints are described in Sect. 4.3. The algorithm maximizes wheel traction or minimizes power consumption depending on the local terrain profile. In highly uneven terrain, traction is optimized. In benign terrain, power consumption is minimized. The algorithm relies on an estimate of wheel-terrain contact angles and a simplified model of the terrain physical properties.

Simulation results are presented in Sect. 4.4 for a planar model of a four-wheeled rover on uneven terrain. It is shown that the rough terrain control method leads to increased traction and improved power consumption compared to traditional individual-wheel velocity control. Experimental results in Sect. 4.4 show that for a six-wheeled rover traversing a ditch the proposed control method increases the net forward wheel thrust and improves rover mobility.

4.2 Mobile Robot Rough Terrain Control (RTC)

Consider an *n*-wheeled vehicle on uneven terrain, as shown in Fig. 4.1. Here it is assumed that the vehicle is skid-steered, and thus only forces in the *XY* plane are considered since out-of-plane forces cannot be actively controlled. (Note that the following analysis can be extended to the case of vehicles with steered wheels.) It is also assumed that each wheel makes contact with the terrain at a single point, denoted $\mathbf{p}_i$, $i = \{1,\ldots,n\}$. Vectors from the points $\mathbf{p}_i$ to the vehicle center of mass $\mathbf{p}_c$ are denoted $\mathbf{V}_i = [V_i^x \; V_i^y]^T$, $i = \{1,\ldots,n\}$ and are expressed in the corresponding local frame $\{xyz_i\}$ at $\mathbf{p}_i$. The 3x1 vector $\mathbf{f}_s$ is expressed in the inertial frame $\{XYZ\}$ and represents the summed effects of vehicle gravitational forces, inertial forces, forces due to manipulation, and forces due to interaction with the environment or other robots.

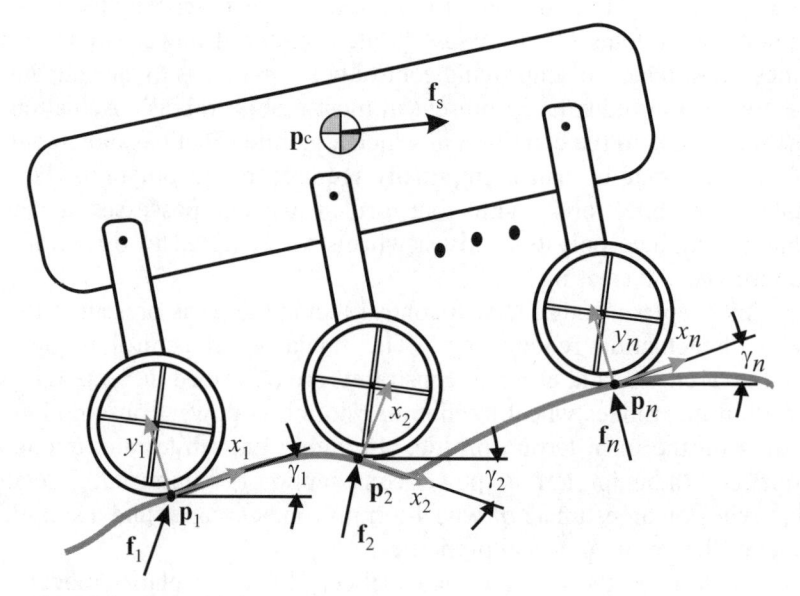

Fig. 4.1. *n*-Wheeled vehicle on uneven terrain

A wheel-terrain contact force exists at each point $\mathbf{p}_i$ and is denoted $\mathbf{f}_i = [T_i \; N_i]^T$, as shown in Fig. 4.2. The vector is expressed in the local frame $\{xyz_i\}$ and can be decomposed into a tractive force T_i tangent to the wheel-terrain contact plane and a normal force N_i normal to the wheel-terrain contact plane. It is assumed that there are no moments acting at the wheel-terrain interface. The angles γ_i, $i = \{1,\ldots,n\}$ represent the angle between the horizontal and the wheel-terrain contact plane.

4.2 Mobile Robot Rough Terrain Control (RTC)

For the planar system above, quasi-static force balance equations can be written as:

$$\begin{bmatrix} {}^0\mathbf{R}_1 & & {}^0\mathbf{R}_2 & & \cdots & & {}^0\mathbf{R}_n & \\ V_1^y & -V_1^x & V_2^y & -V_2^x & \cdots & V_n^y & -V_n^x \end{bmatrix} \begin{bmatrix} \mathbf{f}_1 \\ \vdots \\ \mathbf{f}_n \end{bmatrix} = \begin{bmatrix} F_x \\ F_y \\ M_z \end{bmatrix} \quad (4.1)$$

where ${}^i\mathbf{R}_j$ represents a 2x2 matrix transforming a vector expressed in frame j to one expressed in frame i.

Eq. (4.1) represents the quasi-static force balance on the vehicle and is referred to as the force distribution equations (Hung et al. 1999). It ignores dynamic effects, which are small in low-speed planetary exploration vehicles. Force distribution has been studied for general kinematic chains and legged vehicles (Kumar and Waldron 1988; Kumar and Waldron 1990; Chung and Waldron 1993). Efficient formulation of the force distribution equations for more general vehicles has been addressed (Hung et al. 1999).

Eq. (4.1) can be expressed in matrix form as:

$$\mathbf{Gx} = \mathbf{f}_s \quad (4.2)$$

where the matrix $\mathbf{G}$ is a function the vehicle geometry, wheel-terrain contact locations and wheel-terrain contact angles, $\mathbf{x} = [T_1 \ N_1 \ ... \ T_n \ N_n]^T$, and $\mathbf{f}_s = [F_x \ F_y \ M_z]^T$. In the proposed control approach, $\mathbf{f}_s$ is an input vector. Note that $\mathbf{f}_s$ is directed along the body axis for normal forward driving. Wheel torques are thus sought that result in the desired force vector $\mathbf{f}_s$.

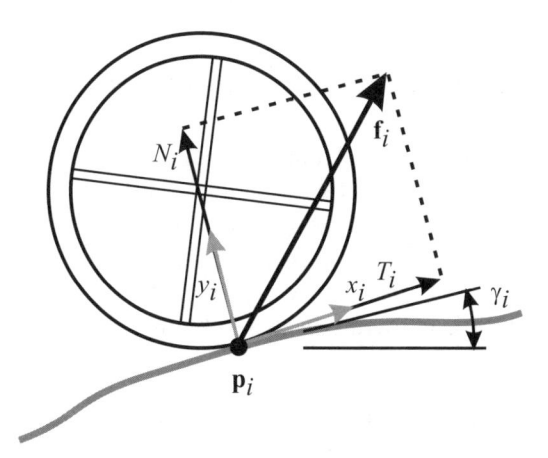

Fig. 4.2. Wheel-terrain interaction force decomposition

Eq. (4.2) represents an underconstrained problem (ignoring the trivial case of a one-wheeled vehicle). There are an infinite number of wheel-terrain contact forces T_i and N_i that balance the vector $\mathbf{f}_s$. In general, a planar system with n wheel-terrain contact points possesses ($2n$-3) degrees of actuation redundancy. The goal of rough terrain control is to find a set of wheel-terrain contact forces (which are modified by means of independently controlled motor torques) that satisfy the force distribution equations and the problem constraints while optimizing an aspect of system performance. The control problem can thus be framed as an optimization problem, and be generally stated as follows: *Optimize system performance subject to the equality constraint* $\mathbf{Gx} = \mathbf{f}_s$ *while satisfying all problem physical constraints.*

A simple approach to solving underconstrained systems is via the pseudoinverse or least-squares solution, defined as $\mathbf{x}^+ = (\mathbf{G}^T\mathbf{G})^{-1}\mathbf{G}^T\mathbf{f}_s$, assuming $(\mathbf{G}^T\mathbf{G})$ is nonsingular. However, while this solution is computationally inexpensive, it does not guarantee satisfaction of the problem's physical constraints. These constraints and an optimization framework for solving Eq. (4.2) are discussed below.

The above analysis has considered a two-dimensional system. The effects of motion on general (i.e. three-dimensional) terrain can be considered by noting that the gravitational component of $\mathbf{f}_s$ is influenced by the vehicle's body roll. Since roll can be easily sensed, the vertical component of $\mathbf{f}_s$ can be modified on-line. This changes the load balance (and thus the available tractive force) on each side of the vehicle. Note that body roll will also influence wheel-terrain longitudinal forces. These effects are not considered here.

4.3 Wheel-Terrain Contact Force Optimization

Mobile robots must maintain adequate wheel traction in highly challenging terrain. However, during travel over benign terrain, it is important for systems to exhibit good power efficiency, since on-board power storage is often limited. Optimization of wheel-terrain contact forces is therefore performed using two criteria: maximum traction or minimum power consumption. These criteria are discussed below.

4.3.1 Optimization Criteria

An optimization criteria for maximizing traction at the wheel-terrain interface can be developed based on the observation that the maximum tractive force a terrain can bear increases with increasing normal force (Bekker 1969). Thus to avoid terrain failure and resulting gross wheel slip, the control algorithm should seek to minimize the maximum ratio of the tractive force to the normal force. An objective function representing this ratio can be written as:

$$R = \max_{i}\left\{\frac{T_i}{N_i}\right\}. \qquad (4.3)$$

Similar criteria have been developed in (Sreenivasan and Wilcox 1994) and an analytical solution to the optimization problem has been developed for a two-wheeled vehicle.

An optimization criteria for minimum power consumption can be developed based on the fact that the power consumed by a DC motor-driven wheeled vehicle using PWM amplifiers can be estimated by the power dissipation in the motor resistances (Dubowsky et al. 1995). Power consumption of the vehicle is related to the motor torques as:

$$P = \frac{Rg^2}{K_t^2}\sum_{i=1}^{n}\Gamma_i^2 \qquad (4.4)$$

where R is the motor resistance, K_t is the motor torque constant, g is the motor gear ratio, and Γ_i is the torque applied by the ith motor. The power consumption can then be related to the tractive force T_i by:

$$P = \frac{Rg^2 r^2}{K_t^2}\sum_{i=1}^{n}T_i^2 \qquad (4.5)$$

where r is the wheel radius. To minimize power consumption the control algorithm should seek to minimize P. Here we have assumed that on benign terrain with little wheel sinkage, the tractive force is the product of the applied wheel torque and the wheel radius. In the case of large wheel sinkage and/or substantial wheel slip, the tractive force (i.e. the net wheel thrust) is a function of the wheel sinkage, slip, and various terrain parameters. However in this case the tractive force still remains proportional to the wheel torque (Wong 1976).

Based on Eqs. (4.3) and (4.5) a dual-criteria objective function that optimizes for maximum traction or minimum power consumption depending on the terrain profile can be developed. In rough terrain, traction should be maximized. In benign terrain, power consumption should be minimized. Terrain roughness can be determined by examining the values of the wheel-terrain contact angles. Consider the switching function S:

$$S = \begin{cases} 1 & \text{if } \max_{i}\{|\gamma_i|\} > C \\ 0 & \text{otherwise} \end{cases} \qquad (4.6)$$

where C is a user-defined threshold level. This function distinguishes between benign and challenging terrain by examining the steepest local terrain slope. Numerous other methods could be used to assess terrain difficulty. For example, a vehicle equipped with a range sensing system could characterize unevenness by analyzing local terrain elevation data.

A dual-criteria objective function which combines Eqs. (4.3), (4.5), and (4.6) can then be expressed as:

$$Q = RS + P(1-S). \tag{4.7}$$

Thus the vehicle force distribution will be optimized for either maximum traction or minimum power consumption, depending on the local terrain profile. The frequency at which Q is evaluated should be a function of the expected spatial frequency of the local terrain.

4.3.2 Problem Constraints

Optimization of the force distribution problem must consider robot physical constraints. One such constraint is that all rover wheels should remain in contact with the terrain. This can be expressed by ensuring that all wheel-terrain normal forces N_i remain positive, or:

$$N_i > 0 \quad \forall i,\ i = \{1,\ldots,n\}. \tag{4.8}$$

The second constraint is that the wheel torques must remain within the saturation limits of the actuator, or:

$$\Gamma_i^{\min} \leq (T_i \cdot r) \leq \Gamma_i^{\max} \quad \forall i,\ i = \{1,\ldots,n\}. \tag{4.9}$$

The third is that the tractive force exerted on the terrain must not exceed the maximum force that the terrain can bear. The simplest approximation of this constraint is a "traction coefficient" model:

$$T_i \leq \mu N_i \quad \forall i,\ i = \{1,\ldots,n\} \tag{4.10}$$

where μ is the wheel-terrain traction coefficient. This approximation is reasonable for rigid wheels traveling over rigid terrain. For rigid wheels in deformable terrain, the maximum shear strength of the terrain can be computed from Coulomb's equation:

$$\tau^{\max} = c + (N_i/A_i)\tan(\phi) \tag{4.11}$$

where A is an estimate of the wheel-terrain contact area. Thus the terrain strength constraint can be written for deformable terrain as:

$$(T_i/A_i) \leq \tau_i^{\max} \quad \forall i,\ i = \{1\ldots n\}. \tag{4.12}$$

A priori estimates of c and ϕ can be used if the terrain properties are known in advance. If the parameters are unknown or variable, estimation

4.4 Results—Rough Terrain Control

4.4.1 Simulation Results

The performance of the multi-criteria rough terrain control algorithm was compared to traditional individual-wheel velocity control in simulation. The simulated system was a planar vehicle with a mass of 10 kg, shown in Fig. 4.3. The rigid wheel radius r was 10 cm and its wheel width b was 15 cm. The wheel spacing l was 0.8 m. Measured quantities were vehicle pitch α and wheel angular velocities ω_1 and ω_2. Sensor noise was modeled by white noise of standard deviation approximately equal to 5% of the full-range values.

The force distribution equations for the system can be written as:

$$\begin{bmatrix} \cos(\gamma_1) & -\sin(\gamma_1) & \cos(\gamma_2) & -\sin(\gamma_2) \\ \sin(\gamma_1) & \cos(\gamma_1) & \sin(\gamma_2) & \cos(\gamma_2) \\ V_1^y & -V_1^x & V_2^y & -V_2^x \end{bmatrix} \begin{bmatrix} T_1 \\ N_1 \\ T_2 \\ N_2 \end{bmatrix} = \begin{bmatrix} F_x \\ F_y \\ M_z \end{bmatrix}. \qquad (4.13)$$

This system of equations possesses $(2n-3)=1$ degree of redundancy. Thus, there exists a free variable that can be selected based on the dual-criteria optimization method discussed above.

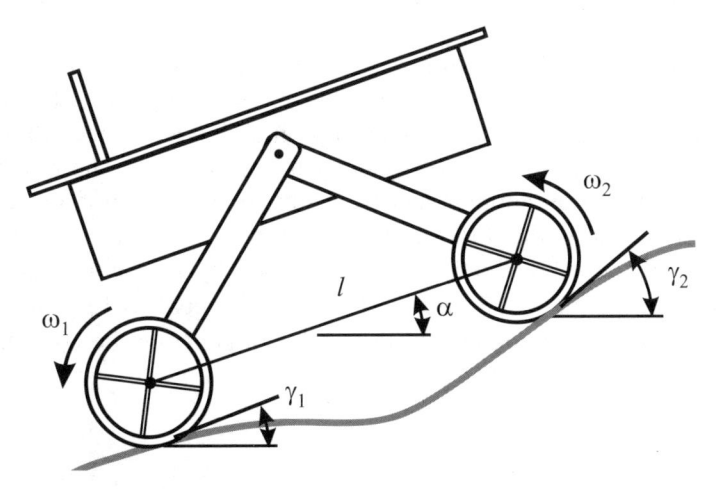

Fig. 4.3. Simulated rover system

The terrain was modeled as a moderately dense deformable soil similar to that which has been observed on Mars (NASA 1988; Matijevic et al. 1997). The following parameters were used:

- Cohesion $c = 1.0$ kPa
- Internal friction angle $\phi = 35°$
- Sinkage coefficient $n = 1$
- Cohesive modulus of deformation $K_c = 10$ kN/m^{n+1}
- Frictional modulus of deformation $K_\phi = 850$ kN/m^{n+2}
- Shear deformation modulus $K = 0.03$ m

At each simulation time increment the wheel sinkage, motion resistance, and wheel thrust were computed as a function of the soil parameters and the applied wheel torque.

Wheel sinkage was computed in order to determine the motion resistance due to soil compaction. Sinkage was computed for each wheel i as (Wong 1976):

$$z_i = \left[\frac{3N_i}{b(3-n)(K_c/b + K_\phi)\sqrt{2r}}\right]^{(2/(2n_s+1))} \qquad (4.14)$$

The motion resistance due to soil compaction was determined by (Wong 1976):

$$R_i = b\left[\left(\frac{K_c}{b} + K_\phi\right)\frac{z_i^{n+1}}{n+1}\right]. \qquad (4.15)$$

The wheel thrust was computed as (Wong 1976):

$$TH = r^2 b \int_0^{\theta_1}\left(c + \left(\left(\frac{k_c}{b}\right)+k_\phi\right)(r(\cos\theta-\cos\theta_1))^n\right)\tan\phi\left(1-e^{-\frac{r}{k}[\theta_1-\theta-(1-i)(\sin\theta_1-\sin\theta)]}\right)\cdot\cos(\theta)d\theta. \qquad (4.16)$$

Two sets of simulation results are presented below. The first simulation was the traverse of gently rolling terrain, as seen in Fig. 4.4. The velocity-controlled system was commanded by an individual-wheel PID control scheme with a desired angular velocity of 2.5 rad/sec. The rough terrain control system was commanded by a horizontal inertial force vector of magnitude equal to the difference between the desired body velocity of 25 cm/sec and the actual body velocity, divided by the vehicle mass. The dual-criteria optimization threshold C was set equal to 15°, since terrain with ground-contact angles less than 15° can generally be considered benign.

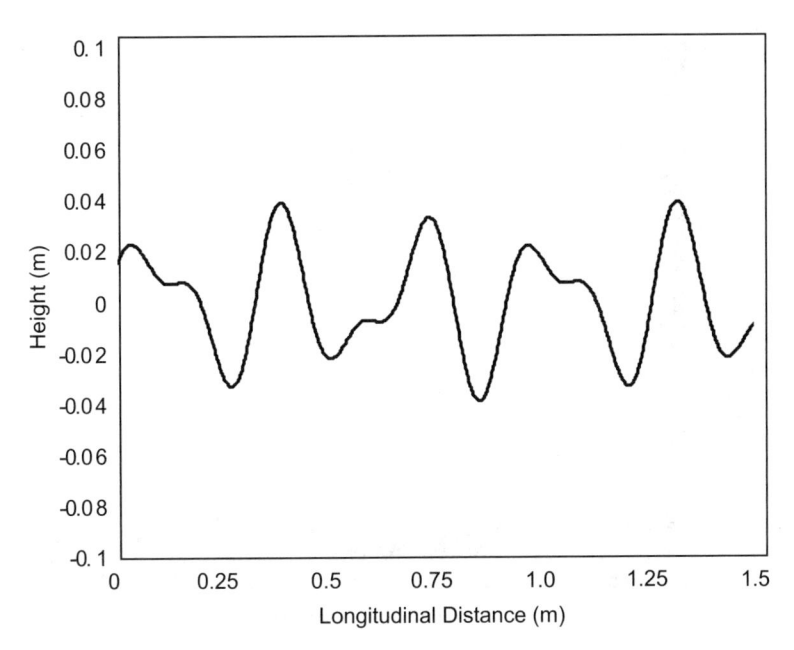

Fig. 4.4. Simulated benign terrain profile

Both the velocity-controlled system and the rough terrain control system successfully traversed the benign terrain. However, the energy consumed by the rough terrain control system was 14.5 J compared to 23.5 J by the velocity-controlled system, an improvement of 38.3%. This power savings is due to reduced wheel slip, as shown in Fig. 4.5. The rough terrain control system has an average slip ratio of 5.3% during the traverse while the velocity controlled system has an average slip of 9.4%. The dual-criteria optimization was in energy-minimization mode during most of the traverse. Thus, even in relatively gentle terrain the rough terrain control can be beneficial by reducing power consumption.

90   Chapter 4   Rough Terrain Control

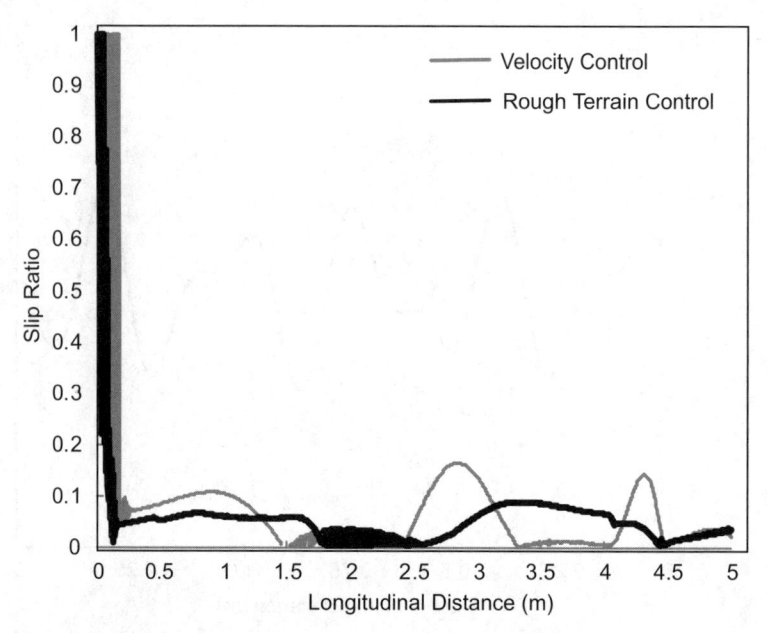

Fig. 4.5. Average slip ratio of front and rear wheels for rough terrain control system and velocity controlled system

The second simulation was the traverse of highly challenging terrain, seen in Fig. 4.6. The maximum slopes in this terrain are near the internal friction angle of the soil. Control parameters were the same as the previous simulation.

In this simulation the rough terrain control system is able to complete the traverse while the velocity-controlled system is not. This is due to the additional thrust force generated by the rough terrain control algorithm, shown in Fig. 4.7. The total wheel thrust generated by the rough terrain control system remains higher than the thrust generated by the velocity-controlled system during most of the traverse. In this case the rough terrain control system commands increased torque to the rear wheel, which has a much higher load than the front wheel, resulting in increased net thrust. The dual-criteria optimization remained in traction maximization mode for the majority of the traverse.

The average wheel slip in the rough terrain control system remained lower than the velocity-controlled system during most of the traverse, as seen in Fig. 4.8. Note that although significant slip remained in the rough terrain control system, this is due to challenging nature of the terrain.

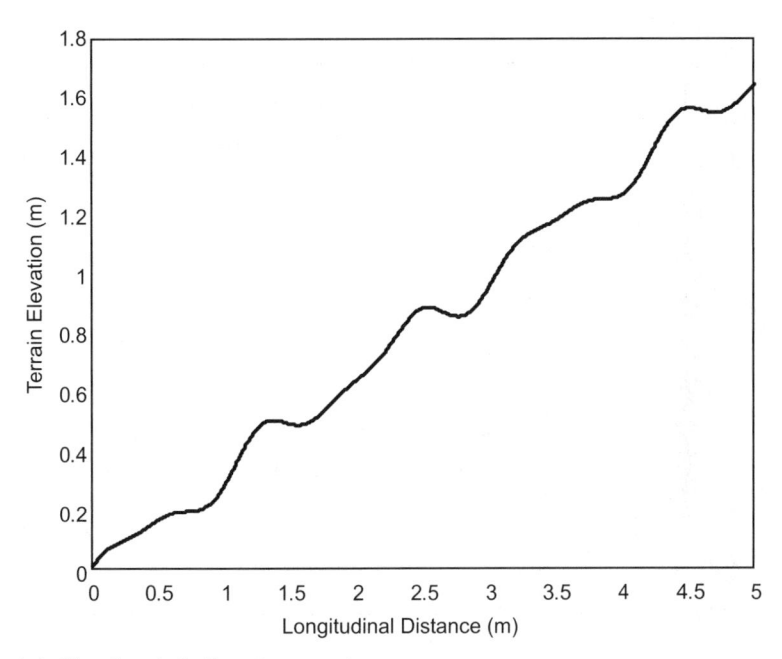

Fig. 4.6. Simulated challenging terrain

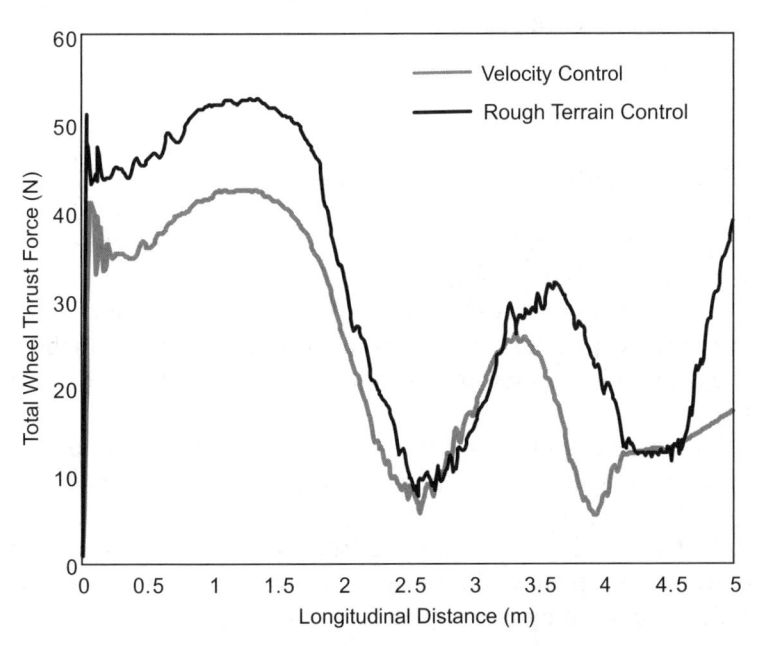

Fig. 4.7. Total wheel thrust of rough terrain control system vs. velocity-controlled system

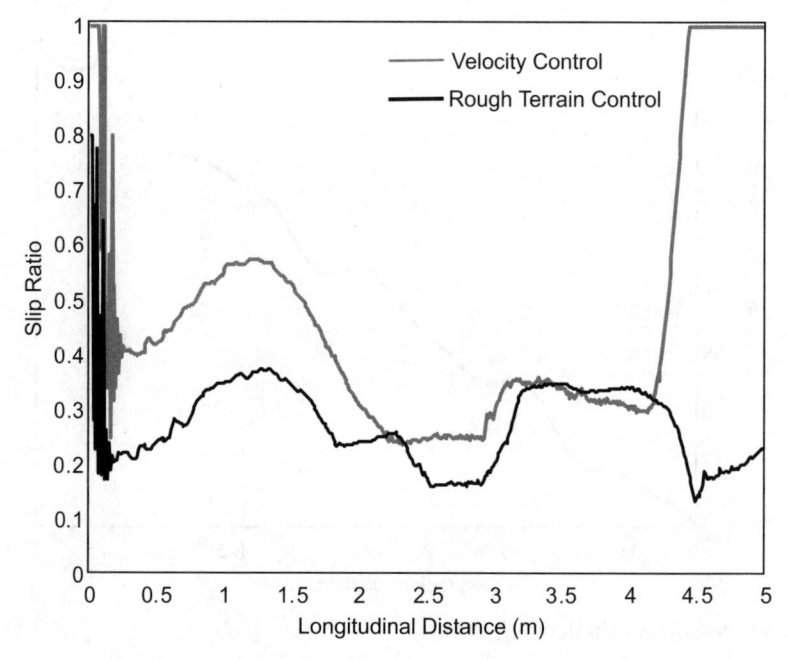

Fig. 4.8. Average slip ratio of front and rear wheels of rough terrain control system vs. velocity-controlled system

4.4.2 Experimental Results

The rough terrain control algorithm was applied to a six-wheeled experimental rover operating in an indoor rough terrain environment, shown in Fig. 4.9. This rover is described in detail in Sect. 2.6.2. First, a go/no-go ditch traversal experiment was performed to examine the mobility improvement provided by the rough terrain control algorithm compared to individual-wheel velocity control.

The rover was commanded to traverse a ditch covered by loose, sandy soil. The maximum depth of the ditch was approximately one wheel diameter. The width of the ditch varied from approximately two to four wheel diameters. The wheel-terrain contact angles were observed to vary greatly during traversal of the ditch, as shown in Fig. 4.10. Thus, ditch traversal is a challenging mobility task.

4.4 Results—Rough Terrain Control 93

Fig. 4.9. Experimental rover during go/no-go ditch traversal experiment

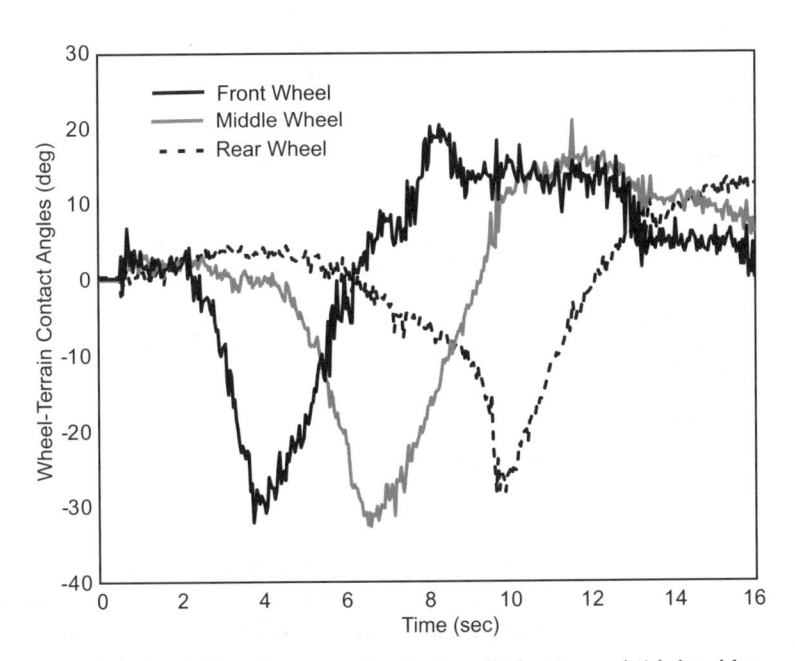

Fig. 4.10. Wheel-terrain contact angles during ditch traversal (right side contact angles shown)

It was observed that the velocity controlled system successfully completed the traverse 6 times out of 20, while the rough terrain control system successfully traversed the ditch 14 times out of 20, an improvement of 133%. Variability in the results are due to irregularity in soil compactness and distribution, and in the ditch traversal route.

Mobility improvement due to rough terrain control can be understood by examining a time history of the rover's right-side wheel normal forces during ditch traversal, shown in Fig 4.11. At time $t = 0$, when the rover is on flat terrain, it can be seen that the system weight is unevenly distributed, with the rear wheel bearing approximately 49% of the rover weight, the middle wheel bearing 35%, and the front wheel 16%. This is due to the robot's suspension configuration. During traversal, the normal forces vary by as much as 87% compared to their initial values.

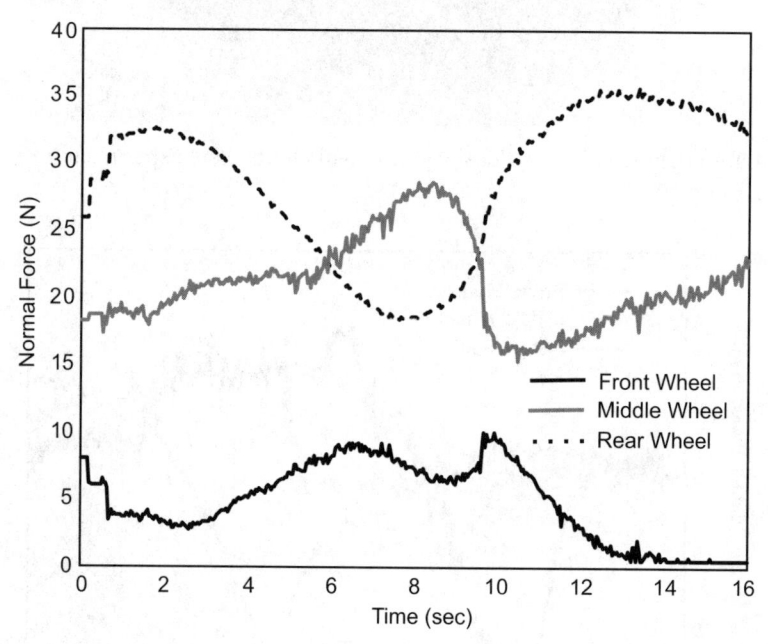

Fig. 4.11. Normal forces during ditch traversal (right side forces shown)

The velocity controller applies the torque necessary to achieve a desired angular velocity. This results in applied thrust that is generally either less than or greater than the maximum thrust the soil can bear. If the applied thrust is less than maximum, the resulting total thrust exerted by the rover is sub-optimal. If the applied thrust is greater than maximum, soil failure occurs, and wheel slip results.

Conversely, the rough terrain control system attempts to apply the maximum thrust the soil can bear. Thus, the rear wheel (which has high normal force) is commanded greater torque than the front wheel (which has a low normal force). The resulting net vehicle thrust is greater than the velocity-controlled system, resulting in improved rough terrain mobility. In these experiments, estimates of the terrain parameters c and ϕ were based on results obtained from parameter identification experiments.

A second experiment was performed to quantify the thrust increase generated by the rough terrain control algorithm. A weighted aluminum sled was attached to a force/torque sensor mounted at the front of the rover, as shown in Fig. 4.12. The force exerted on the sled was measured during the ditch traverse with a six-axis force/torque sensor. Results of a representative pair of trials are shown in Fig. 4.13.

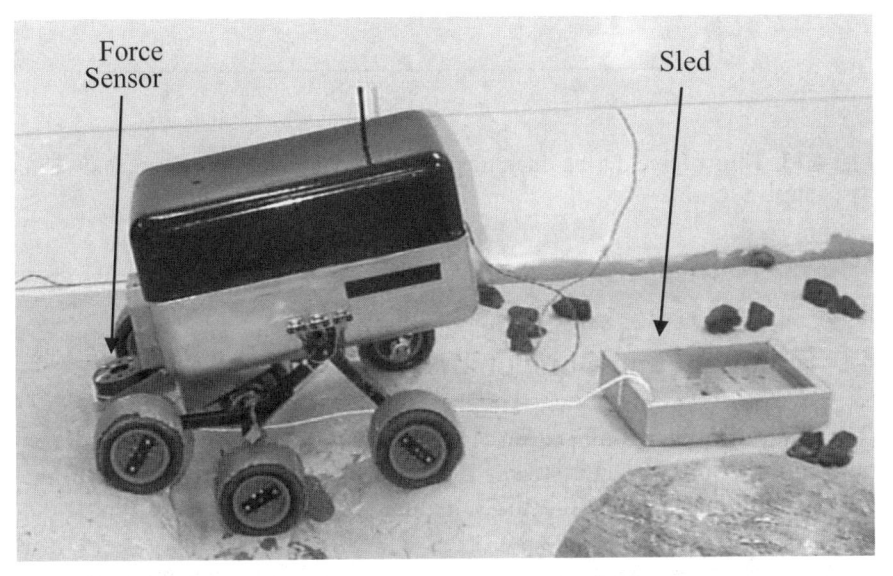

Fig. 4.12. Rover during thrust force measurement experiment

It can be seen that the rough terrain control system generated greater thrust than the velocity-controlled system during the majority of the traverse. Again, this thrust increase is due to optimization of the wheel-torque distribution by the rough terrain control algorithm. The average thrust improvement was 82%, a substantial improvement. This thrust improvement allows a rough terrain control rover to traverse more challenging terrain than a velocity-controlled rover.

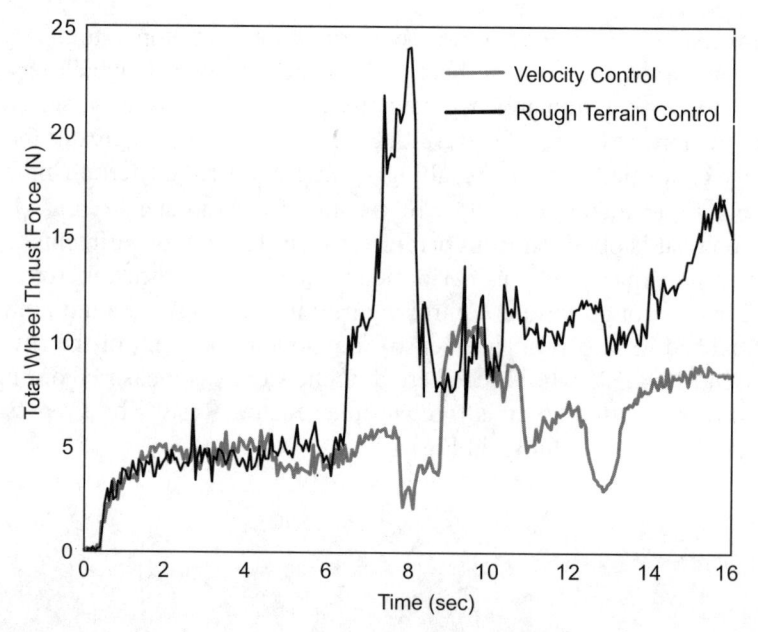

Fig. 4.13. Thrust force during ditch traversal with rough terrain control and velocity control

4.5 Summary and Conclusions

Rough terrain control algorithms must adapt to changing terrain unevenness and traction characteristics to optimize mobility. One approach to accomplishing this is to exploit the actuator redundancy present in most mobile robots.

In this chapter a rough terrain control method has been presented that maximizes traction or minimizes power consumption depending on the local terrain profile. Simulation results have shown that the control algorithm consumes less power and provides greater mobility than traditional individual-wheel velocity control. Experimental results on a six-wheeled rover have shown that the rough terrain control method results in measurably improved mobility over challenging terrain.

Chapter 5
Conclusions and Suggestions for Future Work

5.1 Contributions of This Monograph

This monograph has presented estimation, motion planning, and control methods for improved mobile robot mobility in rough terrain environments, through the use of physical models of the rover and terrain.

In Chap. 1, the rough terrain modeling, motion planning, and control problems were introduced. Important applications were described that require robots to operate in rough, unstructured environments. However, robot mobility in rough terrain is fundamentally different than in structured, indoor environments. Specifically, operation in rough terrain requires that a robot utilize some knowledge of the surrounding terrain. Methods for robot modeling, motion planning, and control were introduced and previous research in these areas was reviewed.

In Chap. 2, a method for estimating terrain cohesion and internal friction angle during rover motion was presented. These parameters allow prediction of terrain traversability, and are important elements of the rough terrain motion planning and control algorithms presented later in the monograph. An algorithm for on-line estimation of wheel-terrain contact angles was also presented. These contact angles are required for rover analysis and control. Simulation and experimental results showed that terrain parameters and wheel-terrain contact angles can be estimated on-line with good accuracy.

In Chap. 3, two rough terrain motion planning algorithms were presented that utilize the models and techniques presented in Chap. 2. A path planning algorithm was presented that allows rapid, autonomous determination of a safe path through uneven terrain. Simulation results showed that the algorithm found shorter, safer paths than a traditional planning method. An algorithm for configuration planning of a rover's actively articulated suspension was also presented. This algorithm was applied to the particular case of improving tipover stability in rough terrain. Simulation and experimental results showed that the algorithm substantially improved the tipover stability of the JPL SRR in rough, natural terrain.

In Chap. 4, a rough terrain control algorithm was presented that uses a multi-criteria optimization method to maximize power efficiency or wheel traction, depending on the local terrain profile. The algorithm relies on the terrain models and estimation techniques presented in Chap. 2. Simulation and experimental results showed that the algorithm leads to improved performance in rough terrain, compared to traditional individual-wheel velocity control.

In summary, this work has presented novel techniques for estimation, motion planning, and control that exploit an understanding of terrain interaction to allow a mobile robot to improve its mobility in rough, unstructured environments.

5.2 Suggestions for Further Work

This monograph has investigated several aspects related to rough terrain mobility. Although substantial work has been completed, numerous avenues of future research exist.

The terrain parameter estimation methodology presented in Chap. 2 is valid for the case of rigid wheels on deformable terrain. Although this is a commonly-occurring case (and is the expected case for planetary exploration rovers) it would be useful to extend the proposed approach to the case of deformable wheels in rigid or deformable terrain. Also, the integration of additional sensing modes (such color and texture) could yield additional cues to improve an overall understanding of terrain composition. This is an area of current research at MIT. Other current work is focused on developing non-parametric methods for traversability assessment. For example, instead of explicitly estimating cohesion and internal friction angle, we are studying methods that result in linguistic (i.e. "poor," "adequate," or "good") assessments of a region's traversability.

The rough terrain planning algorithm presented in Chap. 3 is suitable for on-line implementation, but can be computationally cumbersome for large terrain maps. More efficient search algorithms could be employed in the rapid search step to reduce computation time. Also, the algorithm globally re-plans the path when it is deemed unsafe at any point. Local re-planning could lead to more efficient computation. The integration of recent, highly-efficient search algorithms such as probabilistic roadmaps might be useful in this regard.

The articulated suspension configuration planning algorithm presented in Chap. 3 is a general algorithm with wide applicability, but has been studied for the specific case of tipover stability optimization. This method could be applied to other criteria, such as maximization of wheel traction. It could also be applied to dynamic situations for robots moving at high speeds.

References

Arora J (1989) Introduction to optimum design. McGraw-Hill, New York
Balaram J (2000) Kinematic state estimation for a Mars rover. Robotica 18: 251-262
Bekker G (1956) Theory of land locomotion. University of Michigan Press, Ann Arbor
Bekker G (1969) Introduction to terrain-vehicle systems. University of Michigan Press, Ann Arbor
Bellutta P, Manduchi R, Matthies L, Owens K, Rankin A (2000) Terrain perception for DEMO III. In: Proc Intelligent Vehicle Symposium
Ben Amar F, Bidaud P (1995) Dynamic analysis of off-road vehicles. In: Proceedings of the fourth international symposium on experimental robotics. pp 363-371
Bernard D, Golombek M (1992) Crater and rock hazard modeling for Mars landing. In: Proceedings of the AIAA space conference
Bickler D (1992) A new family of JPL planetary surface vehicles. In: Missions, technologies, and design of planetary mobile vehicles. pp 301-306
Bonnafous D, Lacroix S, Simeon T (2001) Motion generation for a rover on rough terrains. In: Proceedings of the 2001 IEEE/RSJ international conference on intelligent robots and systems. pp 784-789
Borenstein J, Everett B, Feng L (1996) Navigating mobile robots: Systems and techniques. A K Peters Ltd., Wellesley
Brown R, Hwang P (1997) Introduction to random signals and applied Kalman filtering. John Wiley and Sons
Burn R (1997) Design of a Mars exploration rover. M.S. Thesis, Massachusetts Institute of Technology
Carr M (1996) Water on Mars. Oxford University Press
Caurin G, Tschichold-Gurman N (1994) The development of a robot-berrain interaction system for walking machines. In: Proceedings of the IEEE international conference on robotics and automation. vol 2, pp 1013-1018
Chanclou B, Luciani A (1996) Global and local path planning in natural environments by physical modeling. In: Proceedings of the international conference on intelligent robots and systems. pp 1118-1125
Cheok K, Hoogterp F, Fales W, Hobayashi K, Scaccia S (1997) Fuzzy logic approach to traction control design. In: Electronic braking, traction, and stability control. SAE paper number 960957
Cherif M (1999) Kinodynamic motion planning for all-terrain wheeled vehicles. In: Proceedings of the IEEE international conference on robotics and automation. pp 317-322
Cherif M, Laugier C (1994) Dealing with vehicle/terrain interactions when planning the motions of a rover. In: Proceedings of the IEEE international conference on robotics and automation. pp 579-586
Choi B, Sreenivasan S (1998) Motion planning of a wheeled mobile robot with slip-free capability on a smooth uneven surface. In: Proceedings of the IEEE international conference on robotics and automation. pp 3727-3732

References

Chung W, Waldron K (1993) Force distribution by optimizing angles for multifinger systems. In: Proceedings of the IEEE international conference on robotics and automation. vol 3, pp 717-722

Chottiner J (1992) Simulation of a six-wheeled Martian rover called the rocker bogie. M.S. Thesis, Ohio State University

Cunningham J, Corke P, Durrant-Whyte H, Dalziel M (1999) Automated LHDs and underground haulage trucks. Australian J Mining, pp51-53

Dubowsky S, Moore C, Sunada C (1995) On the design and task planning of power-efficient field robotic systems. In: Proceedings of the sixth ANS robotics and remote systems conference

Eckhardt H (1989) Kinematic design of machines and mechanisms. McGraw-Hill, New York

Farritor S (1998a) On modular design and planning for field robotic systems. Ph.D. Thesis, Massachusetts Institute of Technology

Farritor S, Hacot H, Dubowsky S (1998b) Physics-based planning for planetary exploration. In: Proceedings of the IEEE international conference on robotics and automation. pp 278-83

Gajjar B, Johnson R (2002) Kinematic modeling of terrain adapting wheeled mobile robot for Mars exploration. In: Proceedings of the third international workshop on robot motion and control. pp 291-296

Gifford K, Murphy R (1996) Incorporating terrain uncertainties in autonomous vehicle path planning. In: Proceedings of the international conference on intelligent robots and systems. pp 1134-1140

Golub G, van Loan C (1996) Matrix computations. JHU Press

Gonthier Y, Papadopoulos E (1998) On the development of a real-time simulator for an electro-hydraulic forestry machine. In: Proceedings of IEEE international conference on robotics and automation

Hacot H (1998) Analysis and traction control of a rocker-bogie planetary rover. M.S. Thesis, Massachusetts Institute of Technology

Haddad H, Khatib M, Lacroix S, Chatila R (1998) Reactive navigation in outdoor environments using potential fields. In: Proceedings of IEEE international conference on robotics and automation. pp 1232-1236

Hait A, Simeon T (1996) Motion planning on rough terrain for an articulated vehicle in presence of uncertainties. In: Proceedings of the IEEE/RSJ international symposium on intelligent robots and systems. pp 1126-1133

Hayati S, Volpe R, Backes P, Balaram J, Welch W (1996) Microrover research for exploration of Mars. In: Proceedings of the AIAA forum on advanced developments in space robotics

Hebert M, Krotkov E (1992) 3D Measurements from imaging laser radars. Image and vision computing 10(3):170-178

Hung M, Orin D, Waldron K (1999) Force distribution equations for general tree-structured robotic mechanisms with a mobile base. In: Proceedings of IEEE international conference on robotics and automation. pp 2711-2716

Huntsberger T, Baumgartner E, Aghazarian H, Cheng Y, Schenker P, Leger P, Iagnemma K, Dubowsky S (1999) Sensor fused autonomous guidance of a mobile robot and applications to Mars sample return operations. In: Proceedings of the SPIE symposium on sensor fusion and decentralized control in robotic systems. vol 3839

Iagnemma K, Burn R, Wilhelm E, Dubowsky S (1999b) Experimental validation of physics-based planning and control algorithms for planetary robotic rovers. In: Proceedings of the sixth international symposium on experimental robotics

Iagnemma K, Dubowsky S (2000a) Vehicle wheel-ground contact angle wstimation: with application to mobile robot traction control. In: Proceedings of the international symposium on advances in robot kinematics

Iagnemma K, Dubowsky S (2000b) Mobile robot rough-terrain control (RTC) for planetary exploration. In: Proceedings of the 26th ASME biennial mechanisms and robotics conference DETC

Iagnemma K, Kang S, Brooks C, Dubowsky S (2003) Multi-sensor terrain estimation for planetary rovers. In: Proceedings of the international symposium on artificial intelligence, robotics, and automation in space

Iagnemma K, Rzepniewski A, Dubowsky S, Huntsberger T, Pirjanian P, Schenker P (2000c) Mobile robot kinematic reconfigurability for rough terrain. In: Proceedings of the SPIE symposium on sensor fusion and decentralized control in robotic systems

Iagnemma K, Rzepniewski A, Dubowsky S, Schenker P (2003) Control of robotic vehicles with actively articulated suspensions in rough terrain. J Autonomous Robots 14(1)

Iagnemma K, Shibly H, Dubowsky S (2002) On-line terrain parameter estimation for planetary rovers. In: Proceedings of IEEE international conference on robotics and automation

Kang S (2003) Terrain parameter estimation and traversability assessment for mobile robots. M.S. Thesis, Massachusetts Institute of Technology

Kawakami A, Torii A, Motomura K, Hirose S, (2002) SMC rover: planetary rover with transformable wheels. In: Proceedings of the sixth international symposium on experimental robotics. pp 157-162

Kelly A, Stentz A (1998) Rough terrain autonomous mobility—part 2: an active vision predictive control approach. J Autonomous Robots 5: 163-198

Khatib O (1986) Real-time obstacle avoidance for manipulators and mobile robots. International J Robotics Research 5(1)

Kimura H, Hirose H (2002) Development of Genbu: active wheel passive joint articulated mobile robot. In: Proceedings of the international conference on intelligent robots and systems

Kubota T, Kuroda Y, Kunji Y, Yoshimitsu T (2001) Path planning for newly developed microrover. In: Proceedings of the international conference on robotics and automation. pp 3710-3715

Kumar V, Gardner J Kinematics of Redundantly Actuated Closed Chains IEEE Transactions on Robotics and Automation vol 6 Number 2 pp 269-274 1990

Kumar V, Waldron K Force Distribution in Closed Kinematic Chains IEEE Transactions on Robotics and Automation vol 4 Number 6 pp 657-644 1988

Kumar V, Waldron K (1989) Actively coordinated vehicle systems. ASME J mechanisms, transmissions, and automation in design 111: 223-231

Kumar V, Waldron K (1990) Force distribution in walking vehicles. ASME J Mechanical Design 112: 90-99

Latombe JC (1991) Robot motion planning. Kluwer Academic Publishers

Laubach S, Burdick J (1999) An autonomous sensor-based path planner for microrovers. In: Proceedings of the IEEE international conference on robotics and automation. pp 347-354

Laubach S, Burdick J, Matthies L (1998) An autonomous path planner implemented on the Rocky7 prototype microrover. In: Proceedings of the IEEE international conference on robotics and automation. pp 292-297

Laumond J (1998) Robot motion planning and control. Springer-Verlag

Le A, Rye D, Durrant-Whyte H (1997) Estimation of track-soil interactions for autonomous tracked vehicles. In: Proceedings of the IEEE international conference on robotics and automation. pp 1388-1393

Lee H, Tomizuka M (1996) Adaptive vehicle traction force control for intelligent vehicle highway systems (IVHS). In: Proceedings of the ASME conference on dynamics systems and control. pp 17-24

Lee T, Wu C (2003) Fuzzy motion planning of mobile robots in unknown environments. J intelligent and robotic systems 37: 177-191

Linderman R, Eisen H (1992) Mobility analysis simulation and scale model testing for the design of wheeled planetary rovers. In; Missions, technologies and sesign of planetary mobile vehicles. pp 531-537

Lindgren D, Hague T, Probert-Smith P, Marchant J (2002) Relating torque and slip in an odometric model for an autonomous agricultural vehicle. J Autonomous Robots 13: 73-86

Mae Y, Yoshida A, Arai T, Inoue K, Miyawaki K Adachi H (2000) Application of locomotive robot to rescue tasks. In: Proceedings of the 2000 IEEE/RSJ international conference on intelligent robots and systems. pp 2083-2088

Matijevic J, Crisp J, Bickler D, Banes R, Cooper B, Eisen H, Gensler J, Haldemann A, Hartman F, Jewett K, Matthies L, Laubach S, Mishkin A, Morrison J, Nguyen T, Sirota A, Stone H, Stride S, Sword L, Tarsala J, Thompson A, Wallace M, Welch R, Wellman E, Wilcox B (1997) Characterization of Martian surface deposits by the Mars pathfinder rover sojourner. Science 278(5): 1765-1768

Matthies L, Grandjean P (1994) Stochastic performance modeling and evaluation of obstacle detectability with imaging range sensors. In: IEEE transactions on Robotics and Automation 10(6): 783-92

Matthies L, Gat E, Harrison R, Wilcox B, Volpe R, Litwin T (1995) Mars microrover navigation: performance evaluation and enhancement. J Autonomous Robots 2(4)

Mauer G (1995) A fuzzy logic controller for an ABS braking system. In: IEEE transactions on fuzzy systems 3(4): 381-387

Milesi-Bellier C, Laugier C, Faverjon B (1993) A kinematic simulator for motion planning of a mobile robot on a terrain. In: Proceedings of the international conference on intelligent robots and systems. pp 339-343

Mohan S, Williams R (1997) A survey of 4WD traction control strategies. In: Electronic braking, traction, and stability control. SAE paper number 952644

Moore H, Hutton R, Scott R, Spitzer C, Shorthill R (1977) Surface materials of the Viking landing sites on Mars. J Geophysical Research 82(28)

Moore H, Bickler D, Crisp J, Eisen H, Gensler J, Haldemann A, Matijevic J, Reid L (1999) Soil-like deposits observed by Sojourner the Pathfinder rover. J Geophysical Research 104(E4)

Muir P (1987) Kinematic modeling of wheeled mobile robots. J Robotic Systems 4(2): 281-333

Mutambara A, Durrant-Whyte H (2000) Estimation and control for a modular wheeled mobile robot. IEEE Transactions on Control Systems Technology 8(1)

NASA Technical Memorandum 100470 (1988) Environment of Mars 1988. Lyndon B Johnson Space Center

Nilsson N (1980) Principles of artificial intelligence. Tioga Publishing Company

Nohse Y, Hashuguchi K, Ueno M, Shikanai T, Izumi H, Koyama F (1991) A measurement of basic mechanical quantities of off-the-road traveling performance. J Terramechanics 28(4): 359-371

Olin K, Tseng D (1991) Autonomous cross-country navigation. IEEE Expert 6(4): 16-30

Page L, Sanderson A (1996) Robot motion planning for sensor-based control with uncertainties. In: Proceedings of the IEEE international conference on robotics and automation. pp 1333-1340

Olson C, Matthies L, Schoppers M, Maimone M (2001) Stereo ego-motion improvements for robust rover navigation. In: Proceedings of the IEEE international conference On robotics and automation.

Osborn JF (1989) Applications of robotics in hazardous waste management. In: Proceedings of the SME 1989 world conference on robotics research

Pai D, Reissel L-M (1998) Multiresolution rough terrain motion planning. IEEE Transactions on Robotics and Automation 14(1): 19-33

Peynot T, Lacroix S (2003) Enhanced locomotion control for a planetary rover. In: Proceedings of the international conference on intelligent robotics and systems

Papadopoulos E, Rey D (1996) A new measure of tipover stability margin for mobile manipulators. In: Proceedings of the IEEE international conference on robotics and automation.

Pickover C (1995) Generating extraterrestrial terrain. IEEE Computer Graphics and Applications 15(2)

Plackett C (1985) A review of force prediction methods for off-road wheels. J Agricultural Engineering Research 31: 1-29

Reister D, Unseren M (1993) Position and constraint force control of a vehicle with two or more steerable drive wheels. IEEE Transactions on Robotics and Automation 9: 723-731

Sarkar N, Yun X, Kumar V (1994) Control of mechanical systems with rolling constraints. International J Robotics Research 13(1): 55-69

Schenker P, et al. (1997) Lightweight rovers for Mars science exploration and sample return. In: Proceedings of SPIE XVI intelligent robots and computer vision conference. vol 3208: 24-36

Schenker P, Sword L, Ganino A, Bickler D, Hickey G, Brown D, Baumgartner E, Matthies L, Wilcox B, Balch T, Aghazarian H, Garrett M (1997) Lightweight rovers for Mars science exploration and sample return. In: Proceedings of SPIE XVI intelligent robots and computer vision conference. vol 3208: 24-36

Schenker S, Trebi-Ollennu A, Balaram J, Pirjanian P, Huntsberger T, Baumgartner E, Aghazarian H, Kennedy B, Dubowsky S, Apostolopoulos D, Leger P, McKee G (2000) Reconfigurable robots for all-terrain exploration. In: Proceedings of the SPIE symposium on sensor fusion and decentralized control in robotic systems III

Seraji H (1999) Traversability index: a new concept for planetary rovers. In: Proceedings of the IEEE international conference on robotics and automation.

Seraji H, Howard A (2002) Behavior-based robot navigation on challenging terrain: a fuzzy logic approach. IEEE Transactions on Robotics and Automation 18(3)

Shiller Z (2000) Obstacle traversal for space exploration. In: Proceedings of the IEEE international conference on robotics and automation. pp 989-994

Shiller Z, Chen J (1990) Optimal motion planning of autonomous vehicles in 3-dimensional terrains. In: Proceedings of the IEEE international conference on robotics and automation. pp 198-203

Shmulevich I, Ronai D, Wolf D (1996) A new field single wheel tester. J Terramechanics 33(3): 133-141

Siegwart R, Lamon P, Estier T, Lauria M, Piguet R (2002) Innovative design for wheeled locomotion in rough terrain. J Robotics and Autonomous Systems 40: 151-162

Siméon T, Dacre-Wright B (1993) A practical motion planner for all-terrain mobile robots. in: Proceedings of the IEEE international conference on intelligent robots and systems. pp 1357-1363

Spero D, Jarvis R (2002) Path planning for a mobile robot in a rough terrain environment In: Proceedings of the third international workshop on robot motion and control. pp 417-422

Sreenivasan S (1994) Actively coordinated wheeled vehicle systems. Ph.D. Thesis, Ohio State University

Sreenivasan S, Nanua P (1996) Kinematic geometry of wheeled vehicle systems. In: Proceedings of the ASME design engineering technical conferences

Sreenivasan S, Waldron K (1996) Displacement analysis of an actively articulated wheeled vehicle configuration with extensions to motion planning on uneven terrain. Transactions of the ASME J Mechanical Design 118: 312-317

Sreenivasan S, Wilcox B (1994) Stability and traction control of an actively actuated microrover. J Robotic Systems 11(6): 487-502

Stentz A (1994) Optimal and efficient path planning for partially-known environments. In: Proceedings of the IEEE international conference on robotics and automation. pp 3310-3317

Tan H, Chin Y (1991) Vehicle traction control: variable-structure control approach. ASME J Dynamic Systems Measurement and Control 113: 223-230

Tan H, Chin Y (1992) Vehicle Antilock braking and traction control: a theoretical study. International J Systems Science 23(3): 351-365

Tarokh M McDermott G Hayati S, Hung J (1999) Kinematic modeling of a high mobility Mars rover. In: Proceedings of the IEEE international conference on robotics and automation. pp 992-998

Van Zanten A, Erhardt R, Pfaff G (1997) VDC The vehicle dynamics control system of Bosch. In: Electronic braking, traction, and stability control. SAE paper number 950759

Van Zanten A, Erhardt R, Landesfiend K, Pfaff G (1998) VDC Systems development and perspective. In: Electronic braking, traction, and stability control. SAE paper number 980235

Vincent E (1961) Pressure distribution on and flow of sand past a rigid wheel. In: Proceedings of the first international conference on terrain-vehicle systems. pp 859-877

Volpe R (1999) Navigation results from desert field tests of the Rocky 7 Mars Rover prototype. International J Robotics Research 18(7)

Volpe R (2003) Rover Functional autonomy development for the Mars mobile science laboratory. In: Proceedings of 2003 IEEE aerospace conference

Warren C (1993) Fast path planning using modified A^* method. In: Proceedings of the IEEE international conference on robotics and automation pp 662-667

Weisbin C, Rodriguez G, Schenker P, Das H, Hayati S, Baumgartner E, Maimone M, Nesnas I, Volpe R (1999) Autonomous rover technology for Mars sample return. In: International symposium on artificial intelligence, robotics, and automation in space

Wilcox B (1994) Non-geometric hazard detection for a Mars microrover. In: Proceedings of the AIAA conference on intelligent robotics in field, factory, service, and space. vol 2

Welch G, Bishop G (1999) An introduction to the Kalman filter. Technical report TR 95-041, Department of computer science, University of North Carolina at Chapel Hill

Wong J (1976) Theory of ground vehicles. John Wiley and Sons

Wong J (1989) Terramechanics and off-road vehicles. Elsevier

Wong J, Reece A (1967) Prediction of rigid wheels performance based on analysis of soil-wheel stresses part I. J Terramechanics 4(1): 81-98

Yoshida K, Hamano H, Watanabe T (2002) Slip-based traction control of a planetary rover. In: Proceedings of the sixth international symposium on experimental robotics

Yahja A, Stentz A, Singh S, Brumitt B (1998) Framed-quadtree path planning for mobile robots operating in sparse environments. In: Proceedings of the IEEE international conference on robotics and automation

Yong R, Fattah E, Skiadas N (1984) Vehicle traction mechanics. Elsevier Science Publishers

Yoshida K, Watanabe T, Mizuno N, Ishigami G (2003) Slip, traction control, and navigation of a lunar rover. In: International symposium on artificial intelligence, robotics, and automation in space

Zhang H, Kumar V, Ostrowski J (1998) Motion planning with uncertainty. In: Proceedings of the IEEE international conference on robotics and automation. pp 638-643

Index

$A*$, 55, 56, 58
articulated suspension configuration
 planning, 72, 74, 77, 79
articulated suspensions, 2, 10
assumptions, 13, 15

bevameter, 36

conjugate-gradient search, 76
Coulomb's equation, 23

experimental rover system, 45
extended Kalman filter (EKF), 42

force analysis, 21
force balance equations, 85, 89

Gaussian filter, 63
Gruebler mobility criterion, 73

hang-up failure, 57

inverse kinematics, 18

kinematic analysis, 18, 59
kinematic validity, 21, 60

literature review, 4
local re-planning, 100
localization uncertainty, 66

manipulation forces, 75
Mars exploration, 2
modeling of mobile robots, 5, 17
motion planning, 3, 8, 53, 54
motion resistance, 90

nonlinear optimization, 76

planetary exploration, 1, 54
power consumption, 87
pressure-sinkage moduli, 26
PWM amplifiers, 87

range map, 56
rapid path search, 55
ridge regression, 32
robot localization, 62
rocker-bogie, 20, 21, 45, 47
rough terrain control, 84

Sample Return Rover, 77
sensing, 25, 31, 32
shear deformation modulus, 26, 31
shear failure experiments, 36
sinkage, 31
sinkage exponent, 26
skid-steering, 58
static stability, 21, 59
stress distributions, 27

terrain characterization testbed, 35
terrain measurement uncertainty, 61, 63, 65
terrain parameters, 5, 24
terrain roughness, 57
terrain traversability, 61
terrain types, 27
thrust force, 97
tipover stability, 74
traversability assessment, 100

tree-structured wheeled mobile robot, 72

underconstrained systems, 86

wheel sinkage, 90
wheel slip, 92
wheel-terrain contact angle, 6, 39
wheel-terrain interaction, 5, 24
Wiener process, 42

Springer Tracts in Advanced Robotics

Edited by B. Siciliano, O. Khatib, and F. Groen
Published Titles:

Vol. 1: Caccavale, F.; Villani, L. (Eds.)
Fault Diagnosis and Fault Tolerance for Mechatronic Systems: Recent Advances
191 p. 2002 [3-540-44159-X]

Vol. 2: Antonelli, G.
Underwater Robots: Motion and Force Control of Vehicle-Manipulator Systems
209 p. 2003 [3-540-00054-2]

Vol. 3: Natale, C.
Interaction Control of Robot Manipulators: Six-degrees-of-freedom Tasks
120 p. 2003 [3-540-00159-X]

Vol. 4: Bicchi, A.; Christensen, H.I.; Prattichizzo, D. (Eds.)
Control Problems in Robotics
296 p. 2003 [3-540-00251-0]

Vol. 5: Siciliano, B.; Dario, P. (Eds.)
Experimental Robotics VIII
685 p. 2003 [3-540-00305-3]

Vol. 6: Jarvis, R.A.; Zelinsky, A. (Eds.)
Robotics Research – The Tenth International Symposium
580 p. 2003 [3-540-00550-1]

Vol. 7: Boissonnat, J.-D.; Burdick, J.; Goldberg, K.; Hutchinson, S. (Eds.)
Algorithmic Foundations of Robotics V
577 p. 2004 [3-540-40476-7]

Vol. 8: Baeten, J.; De Schutter, J.
Integrated Visual Servoing and Force Control
198 p. 2004 [3-540-40475-9]

Vol. 9: Yamane, K.
Simulating and Generating Motions of Human Figures
176 p. 2004 [3-540-20317-6]

Vol. 10: Siciliano, B.; De Luca, A.; Melchiorri, C.; Casalino, G. (Eds.)
Advances in Control of Articulated and Mobile Robots
259 p. 2004 [3-540-20783-X]

Vol. 11: Kim, J.-H.; Kim, D.-H.; Kim, Y.-J.; Seow, K.-T.
Soccer Robotics
353 p. 2004 [3-540-21859-9]

Printing: Saladruck, Berlin
Binding: Stein+Lehmann, Berlin